P9-CAO-467

This book comes with access to more content online.

Quiz yourself, track your progress,
and score high on test day!

Register your book or ebook at
www.dummies.com/go/getaccess.

Select your product, and then follow the prompts
to validate your purchase.

You'll receive an email with your PIN and instructions.

SAT® Math

2nd Edition

by Mark Zegarelli

A Wiley Brand

SAT® Math For Dummies®, 2nd Edition

Published by: **John Wiley & Sons, Inc.,** 111 River Street, Hoboken, NJ 07030-5774, www.wiley.com

Copyright © 2022 by John Wiley & Sons, Inc., Hoboken, New Jersey

Published simultaneously in Canada

For general information on our other products and services, please contact our Customer Care Department within the U.S. at 877-762-2974, outside the U.S. at 317-572-3993, or fax 317-572-4002. For technical support, please visit https://hub.wiley.com/community/support/dummies.

Wiley publishes in a variety of print and electronic formats and by print-on-demand. Some material included with standard print versions of this book may not be included in e-books or in print-on-demand. If this book refers to media such as a CD or DVD that is not included in the version you purchased, you may download this material at http://booksupport.wiley.com. For more information about Wiley products, visit www.wiley.com.

Library of Congress Control Number: 2021944968

ISBN 978-1-119-82836-5 (pbk); ISBN 978-1-119-82480-082837-2 (ebk); ISBN 978-1-119-82838-9 (ebk)

SKY10030240_100521

Contents at a Glance

Table of Contents

Introduction

Just like going to the senior prom or getting a driver's license, the SAT is one of those milestones in the life of a high school student. I wish I could say it was as much fun as those other things, but if I did, you probably wouldn't believe anything else I say in the rest of the book.

But any way you slice it, the SAT is still there, scheduled for some Saturday morning a few weeks or months from now. Most colleges require you to submit an SAT score as part of your application process. So because there's no getting around it and it's not going away, your best bet is to do some preparation and get the best possible SAT score you can.

That's where this book comes in. The entire book you have in your hot little hands right now is devoted to refining the math skills you need most to succeed on that all-important SATurday.

About This Book

A lot of SAT prep books divide their attention among all three sections of the SAT: English, math, and the writing sample. This is fine as far as it goes, because you probably want to boost all three scores. But in this book, I focus exclusively on math, math, and more math to help you achieve the best score you can on this — what can I say? — most often dreaded part of the test.

The SAT covers a variety of areas, including algebra, geometry, trigonometry, functions and graphs, and statistics and probability. This book focuses on those SAT topics and helps you get used to problem-solving so that you can turn facts and formulas into useful tools.

I wrote this book to give you the best possible advantage at achieving a good score on the math portion of your SAT. There's no shortcut, but most of what you need to work on comes down to four key factors:

>> Knowing the basics inside and out

>> Strengthening SAT-specific math skills

>> Practicing answering SAT questions

>> Time yourself taking SAT practice tests

For that last point, almost every example and problem here is written in SAT format — either as a multiple-choice question or as a student-produced grid-in question. Chapters 3 to 15 contain math skills that are essential to the SAT. And to give you that test-day experience, this book also includes two practice tests, with access to an additional online practice test. That's hundreds and hundreds of questions designed to strengthen your "SAT muscle," so to speak.

This book also has a few conventions to keep in mind:

>> New terms introduced in a chapter, as well as variables, are in *italics*.

>> Keywords in lists and numbered steps are in **boldface.**

>> Any websites appear in `monofont`.

>> For multiple-choice questions, that's a letter from (A) to (D). For grid-in questions, I write the answer as you'd fill it in on the test. So as a test answer, I give $\frac{7}{9}$ as 7/9 or .777 or .778, which are all acceptable ways to write it on your answer sheet.

Foolish Assumptions

This is an SAT prep book, so my first assumption is that you or someone you love (your son or daughter, mom or granddad, or perhaps your cat) is thinking about taking the SAT sometime in the future. If not, you're still welcome to buy the book.

My second assumption is that you're currently taking or have in your life at some point taken an algebra course, even if you feel like it's all a blur. Now, I *wish* I could tell you that algebra isn't very important on the SAT — oh, a mere trifle, hardly a thought. But this would be like saying you can play NFL football without getting rushed at by a bunch of 250-pound guys trying to pulverize you. It just ain't so.

But don't worry — this book is all about the blur and, more importantly, what lies beyond it. Read on, walk through the examples, and then try out the practice problems at the end of each chapter. I can virtually guarantee that if you do this, the stuff will start to make sense.

Icons Used in This Book

In this book, I use these four icons to signal what's most important along the way:

REMEMBER

This icon points out important information that you need to focus on. Make sure you understand this information fully before moving on. You can skim through these icons when reading a chapter to make sure you remember the highlights.

EXAMPLE

Each example is a formal SAT-style question followed by a step-by-step solution. Work through these examples and then refer to them to help you solve the practice problems at the end of the chapter.

TIP

Tips are hints that can help speed you along when answering a question. See whether you find them useful when working on practice problems.

WARNING

This icon flags common mistakes that students make if they're not careful. Take note and proceed with caution!

Beyond the Book

In addition to what you're reading right now, this book comes with a free access-anywhere Cheat Sheet that includes tips to help you prepare for the math sections of the SAT. To get this Cheat Sheet, simply go to www.dummies.com and type **SAT Math For Dummies Cheat Sheet** in the Search box.

You also get access to three full-length online practice tests. To gain access to the online practice, all you have to do is register. Just follow these simple steps:

1. **Register your book or ebook at Dummies.com to get your PIN (go to** www.dummies.com/go/getaccess**).**

2. **Select your product from the drop-down list on that page.**

3. **Follow the prompts to validate your product, and then check your email for a confirmation message that includes your PIN and instructions for logging in.**

If you do not receive this email within two hours, please check your spam folder before contacting the Technical Support website at http://support.wiley.com or by phone at 877-762-2974.

Now you're ready to go! You can come back to the practice material as often as you want — simply log on with the username and password you created during your initial login. No need to enter the access code a second time.

Your registration is good for one year from the day you activate your PIN.

Where to Go from Here

This book is organized so that you can safely jump around and dip into every chapter in whatever order you like. You can strengthen skills you feel confident in or work on those that need some attention.

If this is your first introduction to SAT math, I strongly recommend that you start out by reading Chapter 1. There, you find some simple but vital SAT-specific information that you need to know before you sit down with pencil in hand to take the test.

If you'd like to start out by getting a sense of how ready you are for the SAT, skip forward to Chapter 16 and take a practice test. When you're done, check out Chapter 17 to see how many questions you got correct, and to read through the answer explanation for every question you got wrong.

However, if it's been a while since you've taken a math course, read Chapter 2, which covers the pre-algebra math-skills you need to know before you proceed to the more difficult math later in the book. Chapters 3 and 4, which cover algebra expressions and equations, can get your math brain moving again, and you may find that a lot of this stuff looks familiar as you go along.

Finally, if you read through a few chapters and feel that the book is moving more quickly than you'd like, go ahead and pick up my earlier book, *Basic Math & Pre-Algebra For Dummies* (John Wiley & Sons). There, I adopt a more leisurely pace and spend more time filling in any gaps in understanding you may find along the way.

1
Getting Started with SAT Math

IN THIS CHAPTER

» Knowing how the SAT math test is organized

» Getting familiar with the Reference formulas provided on the first page of every math test

» Understanding how to enter answers to gridded-response questions

» Identifying the math topics tested on the SAT

» Knowing some basic SAT math strategies

» Coming up with a basic plan of action for getting the SAT math score you need

Chapter **1**

Welcome to SAT Math

SAT math — what joy, what utter bliss!

Well, all right — back on Earth, you probably have some work to do before you reach this stage. I promise to do everything in my power to make your study time as painless and productive as possible. All I ask is that you trust in yourself: You already know more than you think you do.

If you've taken (or are currently taking) algebra in school, much of this book may seem like review. The task at hand is to focus your work on the skills you need to get the best SAT score you can. So in this chapter, I give you a basic overview of SAT math, including scoring, calculator use, and how to enter gridded-response questions into the grid.

I also give you a bit of essential SAT math strategy that every student needs to know. I encourage you to think about your goal for the next SAT based on the level you're currently working at.

Finally, I present three SAT success stories, in which three very different students who set and reached different SAT goals got into the colleges that they were aiming for.

SAT Math Basics

The SAT is a college readiness test and, in some U.S. states, is now being used as a skills test for high school graduation. It covers two main subject areas: English and math. Each subject area is scored on a 200-to-800 point scale, resulting in a composite SAT score from 400 to 1,600 points. (If you have any older sisters or brothers who took the SAT before 2016, they may recall that the test used to be scored on a 600- to 2,400-point scale, but that's all history now.)

Here's an overview of the two math sections of the current SAT.

» A 25-minute No Calculator section containing the following:

- 15 multiple-choice questions, which require you to choose the right answer from among four choices, (A) through (D)

- 5 gridded-response questions (also called student-produced response questions), which require you to record your answer in a special grid

» A 55-minute Calculator section containing the following:

- 30 multiple-choice questions

- 8 gridded-response questions

That's a total of 58 questions, each of which counts for 1 point on your raw score of correct answers (from 0 to 58). This raw score is converted to a scaled score (from 200 to 800), which becomes your SAT math score.

Using the Reference list of formulas

Every SAT math section (No Calculator and Calculator) includes a handy Reference list of formulas that you can use while taking the test, as shown in Figure 1-1. As you can see, this list includes a variety of geometric formulas for the area and circumference of a circle, the area of a rectangle and triangle, the Pythagorean Theorem, and other favorites.

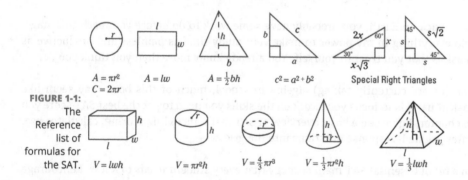

FIGURE 1-1: The Reference list of formulas for the SAT.

$A = \pi r^2$ $A = lw$ $A = \frac{1}{2}bh$ $c^2 = a^2 + b^2$ Special Right Triangles
$C = 2\pi r$

$V = lwh$ $V = \pi r^2 h$ $V = \frac{4}{3}\pi r^3$ $V = \frac{1}{3}\pi r^2 h$ $V = \frac{1}{3}lwh$

TIP

Before taking the SAT, be sure you're familiar with this Reference information, so you'll remember it's there when you're working under time pressure on the actual test.

Filling answers into the grid

Thirteen of the 58 math questions you'll answer when taking the SAT are gridded-response questions. These are Questions 16 to 20 of the No Calculator math test and Questions 31 to 38 of the

Calculator math test. To answer them, you need to fill in the grid provided with your answer sheet.

Most students don't find the grid too difficult to work with. The official SAT instructions for using the grid are provided on every test just before the gridded-response questions. Here are a few things to keep in mind as you work with the grid:

» The grid has four columns, into each of which you can place a digit from 0 to 9, or in some columns a fraction slash (/) or a decimal point (.).

» Be sure to fill in the appropriate bubble in each column so the computer can read your answer. If you don't, your answer will be marked wrong.

» Use the boxes at the top for recording your answer so you can read it easily and change it if necessary.

» The grid contains no minus sign, so all answers are non-negative numbers. (This includes "0".)

» If your answer is a whole number, you can enter it into the grid using any consecutive boxes. For example, here are three equally valid ways to enter the answer "23":

2	3		
	2	3	
		2	3

» If your answer is a fraction, use the fraction slash to record your answer. For example, here are two equally valid ways to enter the answer "1/2".

1	/	2	
	1	/	2

» Don't attempt to enter a mixed number such as $4\frac{1}{2}$ into the grid, because the computer will read this answer as $\frac{41}{2}$ and mark it wrong. Instead, convert mixed numbers to improper fractions or decimals ($4\frac{1}{2} = \frac{9}{2} = 4.5$) and use one of these formats to enter your answer.

» If your answer is a decimal that's less than 1, use the decimal point **without a leading zero** to record your answer. For example, you can enter the answer ".5" in any of following ways:

.	5		
	.	5	
		.	5

» If your answer is a decimal with more decimal places than will fit in the grid, fill in as many decimal places as will fit, either truncating the last decimal place or rounding it up. For example, here are the two ways you can enter the decimal 3.479 into the grid:

3	.	4	7
3	.	4	8

REMEMBER

Although these rules may seem overly complicated, in most cases your actual SAT answers will be relatively straightforward.

Overview of SAT Math Topics

The math that's covered on the SAT is very closely tracked to the math that's covered in most U.S. high school math classes. So if you're a current or recent U.S. high school student, you're probably familiar with most of this curriculum.

The SAT breaks this down into three general areas of study: Heart of Algebra, Problem Solving and Data Analysis, and Passport to Advanced Math (and Other Topics). In this section, I give you an overview of each of these topics.

Heart of Algebra

Heart of Algebra centers on the linear function $y = mx + b$ and other information covered in a typical high school Algebra I class. To answer SAT math questions in this area, you'll need to feel comfortable working with the following:

» Evaluating, simplifying, and factoring algebra expressions (Chapter 3)

» Solving algebra equations and inequalities (Chapter 4)

» Working with linear functions in four complementary ways: words, tables, graphs, and equations (Chapter 5)

» Solving systems of equations (both linear and non-linear), and identifying when such systems have either no solution or infinitely many solutions (Chapter 6)

In Part 2, I cover all these topics in depth.

Problem Solving and Data Analysis

Problem Solving and Data Analysis focuses on a short list of problem-solving techniques:

» Working with ratios, proportional equations, and percentages (Chapter 7)

» Relying on a basic understanding of statistics and probability (Chapter 8)

» Applying these techniques to information presented visually in tables and graphs (Chapter 9)

Part 3 focuses exclusively on these ideas.

Passport to Advanced Math (and Other Topics)

Passport to Advanced Math (and Other Topics) requires you to understand a core of information covered in high school Algebra II:

» Working with functions using $f(x)$ notation, and knowing how to graph a core of basic functions and their most elementary transformations (Chapter 10)

» Understanding how to work with and graph polynomials, especially linear, quadratic, cubic, and quartic polynomials (Chapter 11)

>> Graphing quadratic functions using standard, vertex, and factored forms (Chapter 12)

>> Graphing exponential and radical equations (Chapter 13)

>> Solving problems using basic geometry and trigonometry (Chapter 14)

>> Working with complex numbers and circles on the *xy*-plane (Chapter 15)

I provide this information in Part 4.

SAT Math Strategy

When it comes to doing well on the SAT, your test-taking strategy is a small but important piece of the puzzle. And this strategy also extends to knowing which questions to answer and which to skip, depending on the score you're currently striving for.

In this section, I fill you in on this essential information.

Isn't there some way to get a good SAT Math score without actually knowing math?

No.

I'd love to tell you otherwise, but no.

If the key to getting a great score were, say, choosing Answer C on every question, the name of this book would be Answer C Math For Dummies and it would be a *lot* shorter.

While you fully absorb that difficult truth, I will add that there's a reasonable amount of strategy you should absolutely know before taking your first SAT. And while you may think that lots of students already know this stuff, plenty of others don't — yet.

I don't want you to be one of them.

So read on.

Is there a penalty for guessing?

If you have an older brother or sister who took the SAT before 2016, they may remember the old format, which had a penalty for filling in a wrong answer.

So please take note: **The SAT in its current form has no penalty for filling in a wrong answer.** This goes for all four sections, the English as well as the math sections.

Obviously, then, you want to make sure that you fill in at least some answer for each multiple-choice question on the two math sections. That's 45 questions, so by pure chance, you can expect to get about 11 questions right just by making wild guesses.

Let's take that thinking a step further: If you *don't* fill guesses for all the questions you don't have time to think about, you'll be competing against a ton of other students who *are* guessing. So, bottom line, you can't afford *not* to guess every question you don't know the answer to.

What about the gridded-response questions? Well, because these questions are entirely open ended, you don't have much chance of answering them correctly with a wild guess. But if you have any idea what the answer might be, go ahead and grid it in. Worst case, doing this won't lose you any points.

Are some questions harder than others?

Generally speaking, SAT Math questions fall into three categories of difficulty: easy, medium, and hard.

In both the No Calculator and Calculator sections of the test, the multiple-choice questions are roughly in order from easy, to medium, to hard. And then this pattern repeats, with the shorter grid-in question section also progressing from easy, to medium, to hard.

Table 1-1 shows the rough breakdown of questions by difficulty levels.

TABLE 1-1   Easy, Medium, and Hard Questions

Section 3 — No Calculator			Section 4 — Calculator		
Question Type	Question Number	Difficulty Level	Question Type	Question Number	Difficulty Level
Multiple Choice	1-5	Easy	Multiple Choice	1-10	Easy
	6-10	Medium		11-20	Medium
	11-15	Hard		20-30	Hard
Grid-In	16-17	Easy	Grid-In	31-33	Easy
	18	Medium		34-35	Medium
	19-20	Hard		36-38	Hard

Remember that *every* question counts for one point toward your raw score, which directly affects your scaled score (200–800). So, unlike the tests you take in school, the easiest and hardest questions on the SAT both have the same value.

Do I have to answer every question?

The short answer is, no, you don't have to answer every SAT math question to get a good score.

In fact, depending on your current performance level on practice tests, it may very well be to your benefit *not* to answer all the questions.

This piece of strategy definitely goes against a lot of your training as a high school student. After all, in most of your classes, you can't get an A or even a B on a test without answering just about all the questions. If you only answer 75 percent and skip the rest, even if you answer perfectly, probably the best you can hope for is a C.

However, the situation with the SAT is entirely different.

On the SAT, you can get a 500 math score by answering only about 55 percent of the questions on the test correctly.

I dive more deeply into this aspect of strategy later in this section, when I ask you to consider your own personal starting point, path, and goal for the SAT.

For now — and this goes double if you're a perfectionist — simply let go of the compulsive need to answer all 58 math questions on the SAT. Until you're **already** scoring 740+ on your practice tests, answering all the questions would be a poor allocation of your time. If you're answering all the questions, you're probably rushing through questions that are within your reach, and losing points you should be getting.

So, how many questions should I answer?

The answer to this question depends on your current score, which I break down into three basic scenarios.

Clearing 500

Most schools prefer to enroll students who have an SAT composite score of at least 1,000, which is approximately 500 on both the English and the math tests.

If your English score is 550 or higher, you may be able to get away with a math score that's slightly less than 500. Even so, a good first goal would be to break 500 on the math test.

To get this score, you need to answer about 32 of the 58 SAT math questions correctly. To this end, refer to Table 1-1, and then plan to do the following:

>> Attempt to answer *all* 20 easy questions correctly.

>> Choose 12 out of 18 medium questions to answer correctly, and guess the rest.

>> Guess on the 20 hard questions.

I know it seems weird to guess so many questions. But the SAT is different from the tests you take in school, where you need to get at least 80 percent right to get a decent grade.

Choosing 32 easy and medium questions to focus on gives you almost three minutes per question, which increases your changes of answering more questions correctly. And remember that you have a 25 percent chance of guessing each multiple-choice question correctly, which will also help increase your score a bit.

Believe me, in my experience working with hundreds of students, if you're simply trying to break 500, you probably need to give yourself *more* time by answering *fewer* questions.

Breaking solidly beyond 600

At the next level are students applying for colleges that strongly encourage a composite SAT score of 1,200 or more. That means aiming for at least a 600 score in math, which requires 42 to 45 correct answers.

Here's what I recommend (again, referring to Table 1-1 for question difficulty):

>> Attempt to answer *all* 20 easy questions correctly.

>> Attempt to answer *all* 18 medium questions correctly.

>> Choose 5 to 7 hard questions to attempt to answer, and guess the rest.

In picking the hard questions, aim for familiar-looking problems that you think you've got a shot at answering. Don't spend too much time on any single problem!

Reaching 700 and beyond

If you're striving to break 1,400 or even 1,500 on your SAT composite score, you know that there's no easy answer. You'll want to get a math score of 700 or more, with a little wiggle room if you're confident of scoring 750 or more on the English test.

The good news is that you're obviously a strong student with a well-practiced set of study skills. So, you know that you'll need to at least attempt to answer just about every question on the test, leaving perhaps the two or three hardest questions until the very end.

I recommend getting a private tutor if you don't already have one. Take as many practice tests as you can, and then comb through your wrong answers and do your best to figure out where you went wrong. If your math teacher is supportive, bring especially hard SAT problems to them — they'll almost certainly be willing to help!

When's the latest I can take the SAT and still get into school?

Most students take the SAT with their class in May or June of their junior year. They may try it out before that, but somehow, it doesn't feel real until their whole class is doing it, too.

But if that's the beginning, it doesn't have to be the end.

Usually, December of your senior year is your last shot at the SAT if you want to start college the following fall. Unless you're applying for early acceptance, most colleges don't make their final decisions until after the December SAT scores have been posted.

Hey wait — do I even have to take the SAT to go to college?

I've saved this question for last because if you've read this far, you're clearly an engaged student who's looking for a quality answer to this question rather than an easy one. But in these obviously changing times in education — in both the U.S. and throughout the world — this is a valid question that's worth thinking about.

The short answer is "Definitely maybe."

Even before the COVID-19 pandemic began, a significant number of U.S. colleges and universities had started de-emphasizing the SAT in their entrance requirements and, in some cases, dropping the requirement. And most of them had already begun accepting the ACT in place of the SAT to fulfill this requirement.

The pandemic mostly accelerated these trends.

So a slightly longer answer to the above question would be, "Check the current requirements for the college(s) you're applying to." These requirements may be in flux for the next few years, so stay attuned to any changes as they may be announced.

As for my own humble opinion, I would say that a good SAT (or ACT) score is still likely to open the ivy-draped gates you seek to enter for the foreseeable future. Entrance exams such as these have been around a long, long time. Arguably, they aren't the best statistical indicator of future college success. (That, by the way, would be past success in high school — so keep those grades up!) But large educational institutions — and the institution of education as a whole — tend to be about as responsive and quick to change course as aircraft carriers.

Furthermore, even as colleges weigh the pros and cons of their SAT/ACT requirements, some state school systems have begun requiring the SAT as part of their graduation requirements. And this trend appears to be growing in popularity, at least for the time being.

So to sum up, while your grandchildren may not end up having to take the SAT, if you want to keep your educational options open, you probably won't have the same luxury.

Three SAT Success Stories

Finally, here are three SAT success stories from my recent years of teaching and tutoring. Each of them is a composite of several similar students, with the names changed and all that. But I'm very proud of them all!

Jay's story — clearing 500

When Jay started studying with me, his goal was very straightforward. A gifted lacrosse player, he was already being recruited by a coach at a college where several of his former teammates were already going. They loved it, and he wanted to join them.

So his high school plan was simple: play hard on the lacrosse field, keep his grades up, and break 1,000 on the SAT.

Jay and I worked together for two or three months, and when he took the test for the first time, he got a 520 in English and a 510 in math. With a 1,030 composite SAT score to work with, the coach advocated for him, and Jay received an early acceptance letter a few weeks later.

As much as he enjoyed working with me on SAT math, he was done. As I write this, he's the captain of his college lacrosse team.

Shaun's story — breaking solidly beyond 600

I met Shaun when he was a sophomore, in an SAT class full of all juniors and seniors. He was likeable, quirky, and smart, definitely holding his own in a class full of kids one and two years older than him. His real interest was engineering, and he had a garage full of cool projects in various states of completion.

After the SAT class was done, I started working with him one-on-one over the summer between his sophomore and junior years. He did well on the practice tests, but missed math questions he should have gotten, mostly because he misread the question or made a minor calculation error.

He and I worked together to solidify his skills in the areas of math that most SAT questions focus on: linear functions, linear systems of equations, and quadratic functions. I also encouraged him to spend more time answering the easy-to-medium questions and skipping over the difficult, time-consuming ones.

Shaun thought he was ready to take the SAT for real at the beginning of his junior year. His father thought he needed more practice. I recommended that he take it, if only to resolve their difference of opinion with an actual score. On his first try, he got a 1,340 — 680 in English, 660 in math.

"If you'd like to break 1,400," I suggested, "we can keep on going."

But instead, Shaun just kept his grades up, applied to a good engineering program, and got early acceptance. Case closed.

Amy's story — reaching 700 and beyond

Amy was a bright student, at the top of her class at a very competitive private school. She was already just about killing her SAT practice tests when I began to tutor with her.

We worked together for three or four months, and then she took the SAT with the rest of her class in May of her junior year. On her first try, she got a 750 in English and a 730 in math.

For any other student, that would have been the ballgame. For Amy, getting a 1,480 just about drove her crazy. "Twenty more points! That's all I need!"

We continued through the summer, and she worked tirelessly. For a day or two, just a couple weeks before the August test, I thought she might crack. "You don't have to do this." I explained. "You already have an amazing score. But if you're going to the SAT Olympics, I'm going to coach you at that level."

She pressed on, took the test — and got a 1,530 composite, with a 770 in English and a 760 in math. With her grades, extra-curricular activities, and a tremendous common app essay, she was accepted to her first-choice school. I bet you've heard of it!

IN THIS CHAPTER

» Working with natural numbers, integers, rational numbers, real numbers, and complex numbers

» Reviewing fractions, ratios, decimals, and percentages

» Understanding absolute value

» Calculating with radicals

» Clarifying basic algebra terminology

» Graphing on the *xy*-plane

» Using your calculator on the SAT

Chapter **2**

Review of Pre-Algebra

This chapter provides a review of pre-algebra topics you've probably seen before, but maybe half-remember in a fuzzy sort of way. Although some of these concepts may have given you trouble in 7th or 8th grade, you may be surprised how easy some of this stuff seems now — especially if your current math class is Algebra II or Pre-Calculus!

To begin, I discuss five key sets of numbers: natural numbers, integers, rational numbers, real numbers, and complex numbers. Then, you get a review of four ways to represent rational numbers: as fractions, ratios, decimals, and percentages.

After that, I give you a brief review of absolute value, followed by a more in-depth look at radicals (square roots). Then, I provide a clarification of the algebra vocabulary you may recognize but still be unclear about.

I finish up with a look at a short but important list of calculator moves you'll need to know for the Calculator section of the math SAT.

Sets of Numbers

The SAT Math Test focuses on numbers that fall generally into five sets: natural numbers, integers, rational numbers, real numbers, and complex numbers. Understanding how these sets of numbers differ can be important when answering an SAT question that asks for a solution within a specific set of numbers.

In this section, I give you a brief overview of how these five sets of numbers fit together.

Natural numbers

The first set of numbers you learn as a child are the *natural numbers*, or *counting numbers*, which are the positive whole numbers starting at 1 and continuing without end: $\{1,2,3,4,5,...\}$

When you add or multiply any pair of natural numbers, the result is another natural number. Another way to say this is that the set of natural numbers is closed under both addition and multiplication. However, the natural numbers are not closed under subtraction or division, because when you subtract or divide a pair of natural numbers, the result isn't always a natural number. For example:

$$2-5=-3 \qquad 2 \div 5 = \frac{2}{5}$$

Integers

The next set of numbers are the *integers*, which include the natural numbers, 0, and the negative whole numbers: $\{...,-3,-2,-1,0,1,2,3,..\}$.

The set of integers is closed under addition, subtraction, and multiplication, because when you apply any of these operations to any pair of integers, the result is an integer. However, the integers aren't closed under division, because when you divide a pair of integers, the result isn't always an integer. For example:

$$2 \div 5 = \frac{2}{5}$$

Rational numbers

The *rational numbers* are the set of all numbers that can be expressed as fractions with integers in both the numerator and denominator. For example:

$$\frac{2}{5} \qquad \frac{-14}{3}=-4\frac{2}{3} \qquad \frac{5}{1}=5 \qquad \frac{-11}{1}=-11 \qquad \frac{0}{1}=0$$

As you can see, the set of rational numbers includes all the integers, because every integer can be expressed as the numerator of a fraction with 1 in the denominator.

The set of rational numbers is closed under addition, subtraction, multiplication, and division.

Points on the number line that cannot be expressed as fractions — such as $\sqrt{2}$, $-\sqrt{3}$, and π — are called the set of *irrational numbers*.

Real numbers

The real numbers are the combined set of both rational and irrational numbers. This set includes every point on the number line.

Like the rational numbers, the set of real numbers is closed under the basic four operations. However, the set of real numbers isn't closed under the operation of taking a square root, because the square root of a negative number isn't a real number. For example:

$$\sqrt{-1} = i \qquad \sqrt{-25} = 5\sqrt{-1} = 5i$$

The square root of a negative number is called an *imaginary number* — that is, a real number multiplied by $i = \sqrt{-1}$.

Only a few SAT questions include imaginary numbers. You learn more about imaginary numbers in Chapter 12, where I focus on the roots of quadratic functions, and in Chapter 15, where I discuss operations with imaginary numbers.

Complex numbers

The *complex numbers* are the set of all numbers of the form $a + bi$, where a and b are both real numbers and $i = \sqrt{-1}$. Another way to think of this is that a is a real number and bi is an imaginary number.

The complex numbers include the set of real numbers, the set of imaginary numbers, and other values. Like the rational numbers and real numbers, the complex numbers are closed under the basic four operations. They're also closed under square roots and a variety of other operations.

Very few questions on the SAT require knowledge of the complex numbers. I discuss the specific points you need to know about them in Chapters 12 and 15.

Fractions, Ratios, Decimals, and Percentages

Fractions, ratios, decimals, and percentages are four complementary ways of describing rational numbers — that is, the values that lie between the integers on the number line. In this section, you get a quick review of how to work with these important mathematical values.

Review of fractions and ratios

A fraction is composed of two integers: a *numerator* (top number) divided by a *denominator* (bottom number). For example:

$$\frac{1}{2} \qquad \frac{3}{5} \qquad \frac{9}{10} \qquad \frac{22}{35}$$

The *reciprocal* (or *inverse*) of a fraction is the result when you exchange the numerator and denominator. For example:

$$\frac{2}{1} = 2 \qquad \frac{5}{3} \qquad \frac{10}{9} \qquad \frac{35}{22}$$

Converting between improper fractions and mixed numbers

A *proper fraction* has a numerator that's less than its denominator. In contrast, an *improper fraction* has a numerator that's greater than or equal to its denominator.

Improper fractions can be awkward, because in many real-world cases, an improper fraction doesn't provide easy-to-understand numerical information. For example, if I tell you that I bought $\frac{37}{4}$ gallons of gasoline, you may have a hard time interpreting that information.

However, if I convert the improper fraction $\frac{37}{4}$ to its mixed number of $9\frac{1}{4}$, you now know that I bought a little more than 9 gallons.

TIP

To change an improper fraction to a mixed number, divide the numerator by the denominator. If the result has a remainder, use the remainder as the numerator of the answer. For example, to convert $\frac{26}{3}$ to a mixed number, divide $26 \div 3 = 8$ r 2, so $\frac{26}{3} = 8\frac{2}{3}$.

To convert a mixed number to an improper fraction, multiply the denominator by the whole number, add the numerator, and then use this number as the numerator of the answer. For example, to convert $5\frac{3}{8}$ to an improper fraction, calculate $8 \times 5 + 3 = 43$, so $5\frac{3}{8} = \frac{43}{8}$.

Finding simplified and increased forms of fractions

Sometimes when a fraction has a large numerator and denominator, you can *simplify* it by dividing both of these numbers by the same value, resulting in an equivalent fraction. For example:

$$\frac{3}{6} = \frac{1}{2} \qquad \frac{6}{10} = \frac{3}{5} \qquad \frac{8}{12} = \frac{2}{3} \qquad \frac{20}{25} = \frac{4}{5}$$

The most simplified form of a fraction is usually the form that an SAT answer will take.

You can also reverse this process to *increase the denominator* of a fraction by multiplying the numerator and denominator by the same value.

$$\frac{1}{2} = \frac{4}{8} \qquad \frac{1}{4} = \frac{3}{12} \qquad \frac{2}{5} = \frac{6}{15} \qquad \frac{3}{4} = \frac{18}{24}$$

Increasing the denominator of a fraction is often useful for adding and subtracting fractions, which I discuss in the next section.

Adding and subtracting fractions

When a pair of fractions both have the same denominator, you can add or subtract them by adding or subtracting their numerators and keeping the denominator the same. For example:

$$\frac{1}{5} + \frac{3}{5} = \frac{4}{5} \qquad \frac{6}{7} - \frac{5}{7} = \frac{1}{7} \qquad \frac{5}{9} + \frac{2}{9} = \frac{7}{9} \qquad \frac{11}{13} - \frac{4}{13} = \frac{9}{13}$$

When a pair of fractions have different denominators, you need to get the denominators the same — that is, find a *common denominator* for both fractions. In some cases, you can do this by increasing the lesser denominator to the greater one. For example:

$$\frac{1}{2} + \frac{1}{8} = \frac{4}{8} + \frac{1}{8} = \frac{5}{8} \qquad \frac{7}{10} - \frac{2}{5} = \frac{7}{10} - \frac{4}{10} = \frac{3}{10}$$

In other cases, you need to increase the denominators of both fractions. For example:

$$\frac{1}{2} + \frac{1}{3} = \frac{3}{6} + \frac{2}{6} = \frac{5}{6} \qquad \frac{1}{2} - \frac{2}{7} = \frac{7}{14} - \frac{4}{14} = \frac{3}{14}$$

Multiplying and dividing fractions

Multiplying and dividing fractions is often easier than adding or subtracting them, because you don't need to find a common denominator.

To multiply two or more fractions, first multiply the numerators across to find the numerator of the answer, then multiply the denominators across to find the denominator of the answer. For example:

$$\frac{2}{3} \times \frac{4}{5} = \frac{8}{15} \qquad \frac{5}{6} \times \frac{1}{8} = \frac{5}{48} \qquad \frac{1}{2} \times \frac{3}{4} \times \frac{9}{11} = \frac{27}{88}$$

In some cases, you can make multiplying fractions easier by canceling pairs of equivalent factors in the numerator and denominator. For example, when multiplying $\frac{4}{15}$ and $\frac{5}{8}$, begin by canceling a factor of 4 in both the numerator and denominator:

$$\frac{4}{15} \times \frac{5}{8} = \frac{^1\cancel{4}}{15} \times \frac{5}{^2\cancel{8}}$$

Next, cancel a factor of 5:

$$= \frac{^1\cancel{4}}{^3\cancel{15}} \times \frac{^1\cancel{5}}{^2\cancel{8}}$$

Now when you multiply, the result is already the most simplified form of the answer:

$$= \frac{1}{6}$$

To divide one fraction by another, change the division to multiplication using the mnemonic Keep-Change-Flip: *Keep* the first fraction as it is, *Change* the division sign to a multiplication sign, and *Flip* the second fraction to its reciprocal. For example:

$$\frac{1}{4} \div \frac{3}{5} = \frac{1}{4} \times \frac{5}{3} = \frac{5}{12} \qquad \frac{2}{5} \div \frac{7}{8} = \frac{2}{5} \times \frac{8}{7} = \frac{16}{35} \qquad \frac{3}{10} \div \frac{7}{9} = \frac{3}{10} \times \frac{9}{7} = \frac{27}{70}$$

As when multiplying fractions, you can sometimes make dividing fractions easier by canceling common factors in both the numerator and denominator. Be careful to cancel factors *after* changing the division to multiplication. For example, when dividing $\frac{6}{11}$ by $\frac{2}{3}$, begin by using Keep-Change-Flip:

$$\frac{6}{11} \div \frac{2}{3} = \frac{6}{11} \times \frac{3}{2}$$

Now, cancel a common factor of 2 in both the numerator and denominator:

$$= \frac{^3\cancel{6}}{11} \times \frac{3}{^1\cancel{2}}$$

To complete the problem, multiply:

$$= \frac{9}{11}$$

Connecting fractions and ratios

Once you understand how fractions work, you can easily grasp the related concept of a *ratio*, which is a measurement of the relative sizes of two numbers. For example, if you know that a pet store has a 1:3 (or 1-to-3) ratio of dogs to cats, you know that the number of dogs is $\frac{1}{3}$ the number of cats.

Transforming a ratio to its related fraction in this way is often a good first step when answering SAT questions that involve ratios.

I discuss ratios, plus the related concept of proportional equations, in Chapter 7.

Review of decimals and percentages

Like fractions and ratios, decimals and percentages are two common ways to represent rational numbers that aren't integers. On the SAT, you'll probably need to convert decimals to percentages, and vice versa. In this section, I review this relatively simple process.

Converting decimals to percentages

To change a decimal to a percentage, move the decimal point two places to the right and attach a percent sign. For example:

$$0.45 = 45\% \qquad 1.12 = 112\% \qquad 0.03 = 3\%$$

As you can see from the last example, after changing a decimal to a percentage, you can safely drop unnecessary leading zeros.

In some cases, you may need to include one or more place-holding zeros in the percentage that you create. For example:

$$0.1 = 10\% \qquad 2.5 = 250\% \qquad 10 = 1,000\%$$

When a decimal includes three or more decimal places, the decimal point remains in the resulting percentage. For example:

$$0.234 = 23.4\% \qquad 1.005 = 100.5\% \qquad 1.099 = 109.9\%$$

Changing percentages to decimals

Converting a percentage to a decimal reverses the process I describe in the previous section. To convert a percentage that doesn't have a decimal point, introduce a decimal point just before the percent sign, then move it two places to the left and drop the percent sign. For example:

$$37\% = 0.37 \qquad 175\% = 1.75 \qquad 250\% = 2.5$$

As you can see from this example, after changing a percentage to a decimal, you can safely drop unnecessary trailing zeros.

For percentages that are less than 10%, introduce one or more place-holding zeros as needed into the decimal that you create. For example:

$$5\% = 0.05 \qquad 0.9\% = 0.009 \qquad 0.01\% = 0.0001$$

Absolute Value

The *absolute value* of a number is its non-negative value. For example:

$$|5| = 5 \qquad |-3| = 3 \qquad |0| = 0$$

As you can see, when absolute value is applied to a negative number such as −3, the minus sign is dropped. Otherwise, absolute value has no effect on non-negative numbers such as 5 and 0.

When evaluating an absolute value, evaluate the expression inside the absolute value bars, then apply the absolute value, and then evaluate the rest of the expression. For example:

EXAMPLE

Which of the following is equivalent to $|-2+5|-|-3-7|$?

(A) -13

(B) -7

(C) 7

(D) 13

To begin, simplify the two expressions inside the absolute value bars:

$$|-2+5|-|-3-7| = |3|-|-10|$$

Next, substitute $|3| = 3$ and $|-10| = 10$:

$$= 3-10$$

Now complete the problem:

$$= -7$$

Therefore, Answer B is correct.

Radicals

A *radical* (also called a *root*) is the inverse of an exponent — that is, a radical "undoes" an exponent. For example:

$$3^2 = 9 \text{ so } \sqrt{9} = 3 \qquad 5^3 = 125 \text{ so } \sqrt[3]{125} = 5 \qquad 2^{10} = 1{,}024 \text{ so } \sqrt[10]{1{,}024} = 2$$

The most common type of radical on the SAT is the square root. In this section, I discuss the basics of radicals.

Understanding radicals

Radicals, such as $\sqrt{2}$, arise from reversing the process of squaring a number. That is, they are inverse operations. Table 2-1 shows you the results of squaring the first five positive integers and then taking the square root (radical) of the result.

TABLE 2-1 ## Squaring and Taking a Square Root (Radical) Are Inverse Operations

Squares	Square Roots (Radicals)
$1^2 = 1 \times 1 = 1$	$\sqrt{1} = 1$
$2^2 = 2 \times 2 = 4$	$\sqrt{4} = 2$
$3^2 = 3 \times 3 = 9$	$\sqrt{9} = 3$
$4^2 = 4 \times 4 = 16$	$\sqrt{16} = 4$
$5^2 = 5 \times 5 = 25$	$\sqrt{25} = 5$

When you understand how to place radicals such as $\sqrt{1}$, $\sqrt{4}$, and so forth on the number line, you can see how other values such as $\sqrt{2}$, $\sqrt{3}$, and so on also fit in. Figure 2-1 shows you a number line that includes specified values of radicals.

FIGURE 2-1:
Radicals
on the
number line.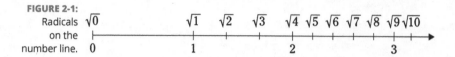

Notice that radicals of square numbers are always equivalent to integers. In contrast, radicals like $\sqrt{2}$ and $\sqrt{3}$ are irrational numbers that fit between these integers on the number line.

When you see this ordered relationship, you can estimate the value of a radical by finding the two integer values where it must fit on the number line. For example:

EXAMPLE

Where on the number line does the value of $\sqrt{39}$ occur?

(A) Between 4 and 5

(B) Between 5 and 6

(C) Between 6 and 7

(D) Between 7 and 8

The number 39 falls on the number line between the square numbers 36 and 49. Therefore, $\sqrt{39}$ falls between $\sqrt{36} = 6$ and $\sqrt{49} = 7$, so Answer C is correct.

Multiplying and dividing radicals

You can multiply any pair of radicals simply by multiplying the values inside the radicals. For example:

$$\sqrt{2}\sqrt{5} = \sqrt{10} \qquad \sqrt{3}\sqrt{11} = \sqrt{33} \qquad \sqrt{6}\sqrt{7} = \sqrt{42}$$

Similarly, you can divide one radical by another by dividing their contents:

$$\frac{\sqrt{14}}{\sqrt{7}} = \sqrt{2} \qquad \frac{\sqrt{30}}{\sqrt{3}} = \sqrt{10} \qquad \frac{\sqrt{55}}{\sqrt{11}} = \sqrt{5}$$

EXAMPLE

$$\frac{\sqrt{15}\sqrt{6}}{\sqrt{10}}$$

If x is equivalent to the expression shown here, what is the value of $2x + 5$?

To find the value of x, first simplify $\frac{\sqrt{15}\sqrt{6}}{\sqrt{10}}$. Begin by multiplying the two radicals in the numerator:

$$\frac{\sqrt{15}\sqrt{6}}{\sqrt{10}} = \frac{\sqrt{90}}{\sqrt{10}}$$

Now, divide the numerator by the denominator and simplify:

$$= \sqrt{9} = 3$$

Thus, $x = 3$, so plug this value into $2x + 5$ and evaluate:

$$2(3) + 5 = 6 + 5 = 11$$

Therefore, the answer is 11.

Simplifying radicals

In some cases, you can simplify a radical by factoring it and then evaluating one of the factors as a whole number. This is always the case when the value inside the radical is divisible by a square number, such as 4, 9, 16, 25, and so forth.

For example, here are three radicals that can be simplified by factoring out $\sqrt{4}$ and then simplifying this value to 2:

$$\sqrt{8} = \sqrt{4}\sqrt{2} = 2\sqrt{2} \qquad \sqrt{12} = \sqrt{4}\sqrt{3} = 2\sqrt{3} \qquad \sqrt{20} = \sqrt{4}\sqrt{5} = 2\sqrt{5}$$

Similarly, you can simplify the following three radicals by factoring out $\sqrt{9}$ and then simplifying this value to 3:

$$\sqrt{18} = \sqrt{9}\sqrt{2} = 3\sqrt{2} \qquad \sqrt{27} = \sqrt{9}\sqrt{3} = 3\sqrt{3} \qquad \sqrt{45} = \sqrt{9}\sqrt{5} = 3\sqrt{5}$$

These three radicals can be simplified by factoring out $\sqrt{16}$:

$$\sqrt{32} = \sqrt{16}\sqrt{2} = 4\sqrt{2} \qquad \sqrt{48} = \sqrt{16}\sqrt{3} = 4\sqrt{3} \qquad \sqrt{80} = \sqrt{16}\sqrt{5} = 4\sqrt{5}$$

And these three can be simplified by factoring out $\sqrt{25}$:

$$\sqrt{50} = \sqrt{25}\sqrt{2} = 5\sqrt{2} \qquad \sqrt{75} = \sqrt{25}\sqrt{3} = 5\sqrt{3} \qquad \sqrt{125} = \sqrt{25}\sqrt{5} = 5\sqrt{5}$$

An SAT question may ask you to simplify a radical. For example:

EXAMPLE

Which of the following is the most simplified form of $\sqrt{72}$?

(A) $2\sqrt{8}$

(B) $3\sqrt{8}$

(C) $6\sqrt{2}$

(D) $6\sqrt{3}$

The number 72 is divisible by 9, so you can factor and simplify $\sqrt{72}$ as follows:

$$\sqrt{72} = \sqrt{9}\sqrt{8} = 3\sqrt{8}$$

Although $3\sqrt{8}$ is equivalent to $\sqrt{72}$, the question asks you to find the most simplified form of the radical. To do this, notice that 8 is divisible by 4, so you can further factor and simplify $3\sqrt{8}$ as follows:

$$3\sqrt{8} = 3\sqrt{4}\sqrt{2} = 3 \cdot 2\sqrt{2} = 6\sqrt{2}$$

This result is the most simplified form of $\sqrt{72}$, so Answer C is correct. (You can also find this answer directly as $\sqrt{72} = \sqrt{36}\sqrt{2} = 6\sqrt{2}$.)

TIP

Other opportunities for simplifying radicals may arise when working with the quadratic formula, which includes a radical. In Chapter 3, I show you how to use this formula to solve quadratic equations. And then in Chapter 12, you use it again to find the roots of quadratic functions.

Adding and subtracting radicals

You can only add equivalent radicals. For example:

$$\sqrt{3} + \sqrt{3} = 2\sqrt{3} \qquad \sqrt{7} + \sqrt{7} + \sqrt{7} = 3\sqrt{7} \qquad \sqrt{2} + \sqrt{2} + \sqrt{2} + \sqrt{2} = 4\sqrt{2}$$

You can add radical expressions that have the same radical parts by adding coefficients. For example:

$$\sqrt{5} + 2\sqrt{5} = 3\sqrt{5} \qquad 3\sqrt{2} + 2\sqrt{2} = 5\sqrt{2} \qquad \sqrt{11} + 4\sqrt{11} + 10\sqrt{11} = 15\sqrt{11}$$

You can also subtract radicals in the same way:

$$5\sqrt{3} - \sqrt{3} = 4\sqrt{3} \qquad 7\sqrt{2} - 4\sqrt{2} = 3\sqrt{2} \qquad 10\sqrt{5} - 2\sqrt{5} - 2\sqrt{5} = 6\sqrt{5}$$

In some cases, you can add or subtract radical expressions with non-equivalent radical parts by simplifying them using the factoring method I describe in the previous section. For example, here's an SAT question that depends upon your understanding this idea:

EXAMPLE

Which of the following is equivalent to $\sqrt{20} + \sqrt{45}$?

(A) $5\sqrt{3}$

(B) $3\sqrt{5}$

(C) $5\sqrt{5}$

(D) $\sqrt{65}$

Begin by factoring $\sqrt{20}$ and $\sqrt{45}$ to simplify them:

$$\sqrt{20} = \sqrt{4}\sqrt{5} = 2\sqrt{5} \qquad \sqrt{45} = \sqrt{9}\sqrt{5} = 3\sqrt{5}$$

Now, rewrite $\sqrt{20} + \sqrt{45}$ and solve it:

$$\sqrt{20} + \sqrt{45} = 2\sqrt{5} + 3\sqrt{5} = 5\sqrt{5}$$

So Answer C is correct.

Rationalizing radicals in the denominator

In some cases, when a radical appears in the denominator of a fraction, an SAT question will require you to *rationalize the denominator* — that is, find an equivalent form of that fraction with an integer in the denominator.

To rationalize the denominator of a fraction, multiply both the numerator and denominator by the radical that's in the denominator. For example, here's how you rationalize $\frac{3}{\sqrt{2}}$:

$$\frac{3}{\sqrt{2}} = \frac{3}{\sqrt{2}} \cdot \frac{\sqrt{2}}{\sqrt{2}} = \frac{3\sqrt{2}}{2}$$

In some cases when rationalizing, you may need to simplify the result. For example, here's how you rationalize $\frac{8\sqrt{5}}{\sqrt{6}}$:

$$\frac{8\sqrt{5}}{\sqrt{6}} = \frac{8\sqrt{5}}{\sqrt{6}} \cdot \frac{\sqrt{6}}{\sqrt{6}} = \frac{8\sqrt{30}}{6}$$

This result can be simplified by dividing both the numerator and the denominator by 2:

$$= \frac{4\sqrt{30}}{3}$$

Understanding Algebra Terminology

Look over the following list of 12 words:

Algebra	Arithmetic	Coefficient	Constant
Equation	Expression	Identity	Inequality
Polynomial	Term	Variable	Zebra

First of all, ignore the word *zebra*. I just put that one in to fill out the last row. Zebras have nothing at all to do with math. They can't even do math.

As for the other 11 words, even though they may get thrown around a lot in whatever math class you're currently taking, you may not be really sure about some (or all!) of them.

You could ask your teacher to explain them. In fact, I encourage you to do that. Even so, you probably won't. After years of teaching, I get that most of my students just don't feel comfortable asking questions about stuff that they think they "should already know."

That's why I'm here: for starters, to help you succeed on the SAT. But beyond this, so that the next time your teacher asks for a volunteer to identify the coefficient of the third term in the right-hand expression of the equation they just wrote on the board, you'll be able to raise your hand with pride and confidence.

Algebra and arithmetic

Arithmetic is the mathematics of numbers and operations upon them — essentially, number crunching. I cover a lot of this material in *Basic Math & Pre-Algebra For Dummies* (Wiley). When you know how to perform the operations in a given problem in the correct order (using the order of operations, affectionately remembered as PEMDAS), you're well on your way to a complete mastery of arithmetic. In fact, you may already be a winner.

In contrast, *algebra* takes arithmetic one step further by introducing the *variable*, which is often named *x*. Algebra is the place that some of the nicest people you will ever meet — possibly, your parents — threw in the towel on math. And then, when they meet me at a party (yes, math teachers sometimes go to parties), they say, "OMG, I could never learn algebra!"

I feel their pain. At the same time, even though using variables can be tricky, there is no law that says arithmetic is always easier than algebra. For example:

Arithmetic Problem	**Algebra Problem**
$$\dfrac{\sqrt{74(17^3-1239)}}{13^3+851(98+2^{20})} - \dfrac{(19^4-1954)(214+8^5)}{\dfrac{\sqrt{864(33+(62^2-793)}}{3^{11}-951+7(2^{20}-4831)}} = \underline{\hspace{1cm}}$$	$x+1=5$

You can probably see that in the algebra problem shown here, the value of *x* is 4. In contrast, the answer to the arithmetic problem is *left for the reader to solve* (which is a fancy way teachers get to pose hard questions without having to answer them!).

The take-away here is that arithmetic is math that involves numbers and operations on them, and algebra includes variables.

Equations, identities, and inequalities

An *equation* is any valid mathematical statement that includes an equals sign (=). For example:

$$x = 17 \qquad 2x - 3 = 6x + 9 \qquad x^2 - 4x + 3 = x - 1$$

One of the key tasks in algebra is to solve equations. Usually, that means discovering the value of the variable in that equation.

In some cases, an equation is true for all (or just about all) values of the variable. An equation like this is called an *identity*. For example:

$$2x + 3x = 5x \qquad x^2 - 1 = (x+1)(x-1) \qquad x^{-1} = \frac{1}{x}$$

The first two identities are true for all values of x. The third is true for all values of x except 0, because a value of 0 in the denominator is not allowed. Identities can be helpful because they allow you to rewrite, and even *rethink*, an equation in a different and potentially more helpful form.

Finally, an *inequality* is a valid mathematical statement that includes one of four *inequality operators*: less than (<), greater than (>), less than or equal to (≤), and greater than or equal to (≥). For example:

$$x < 5 \qquad y + 1 > -7 \qquad 2x - 6 \leq 5x + 12 \qquad x^2 + 2x + 1 \geq 0$$

You can't usually solve an inequality for a specific value, as you can with an equation. Rather, an inequality is usually solved for a *solution set* — that is, a set of values that satisfy the inequality. For example, the solution set for the first inequality is $(-\infty, 5)$, which is the set of numbers less than (that is, up to but not including) 5.

Expressions

An *expression* is any string of mathematical symbols that you can place on one side of an equation, identity, or inequality. For example, consider this equation from the last section:

$$x^2 - 4x + 3 = x - 1$$

This equation includes two expressions: $x^2 - 4x + 3$ and $x - 1$.

As a mathematical concept, expressions often get lost in the shuffle when students are struggling to learn algebra. If you can, try not to let that happen as you revisit algebra while studying for the SAT. Think of it like this:

>> An *expression* doesn't have an equals sign (=). Expressions are evaluated, simplified, or factored, but not solved.

>> An *equation* is two expressions joined with an equals sign. Equations are to be solved.

In Chapter 3, you work extensively with expressions. In this section, I discuss a few more math words that help you to break down and understand expressions better.

Variable and constant terms

Every expression is built from one or more *terms*, separated by either plus signs or minus signs.

Table 2-2 illustrates this concept, showing expressions with from one to four terms.

TABLE 2-2   Expressions with 1, 2, 3, and 4 Terms

Expression	Number of Terms	Terms
$-16x^2yz^3$	1	$-16x^2yz^3$
$x^{10}+1$	2	x^{10} and 1
$4x+3y-5$	3	$4x$, $3y$, and -5
$-x^3-8x^2+x+4$	4	$-x^3$, $-8x^2$, x, and 4

Note that each term includes the sign that precedes it. When no sign is written explicitly, that term is positive; when it includes a minus sign, it's a negative term.

Breaking terms into coefficient and variable

Each term in an expression can be further broken up into coefficients and variables:

>> The *coefficient* of a term is the numerical part, including the minus sign when the term is negative. It's usually listed before the variable, at the beginning of the term.

>> The *variable* part of a term is everything except the coefficient, which includes all variables and their exponents.

Identifying the coefficient and variable parts of a term is useful when you want to simplify an expression by combining like terms (as I explain in Chapter 3).

Table 2-3 shows you how to break the expression $-x^3y^2-8x^2y+x+4$ into four terms, and then break each term further into coefficient and variable.

TABLE 2-3   The Expression $-x^3y^2-8x^2y+x+4$ and Its Coefficients and Variables

Expression	Coefficient	Variable
$-x^3y^2$	-1	x^3y^2
$-8x^2y$	-8	x^2y
x	1	x
4	4	None (Constant term)

Polynomial basics

A polynomial with a single variable (x) is any expression of the following form:

$$a_nx^n + a_{n-1}x^{n-1} + a_{n-2}x^{n-2} + \cdots + a_2x^2 + a_1x + a_0$$

This eye-glazing definition becomes a lot clearer when you see a few sample polynomials:

$$3x^2 + 4x - 5 \qquad -x^7 - 2x^5 - 4x^2 + 8x - 108 \qquad x^{24} - 1$$

As you can see, each term in a polynomial is built from a variable (usually x) raised to any non-negative exponent and then multiplied by a coefficient, with terms either added or subtracted.

For clarity, polynomials are usually written in *standard form*, starting with the term that has the greatest exponent of x and in descending order down to the term that has the least exponent. (Note that the constant term of a polynomial technically has an exponent of 0, so it appears as the last term of a polynomial in standard form.)

Understanding the degree of a polynomial

The *degree* of a polynomial is determined by the exponent of its *leading term* — that is, the first term of the polynomial when arranged in standard form.

The leading term of a polynomial is instrumental in naming the *degree* of that polynomial. Table 2-4 provides examples of the four most common polynomials.

TABLE 2-4 ### Polynomials of Degree 1, 2, 3, and 4

Polynomial	Degree	Name of Degree
$x - 2$	1	Linear
$-x^2 + 3x - 7$	2	Quadratic
$5x^3 + x^2 + 4$	3	Cubic
$x^4 - 1$	4	Quartic

For more detailed information about polynomials, flip to Chapter 11.

Counting the terms of a polynomial

Polynomials are also classified by the number of terms that they have:

>> A *monomial* is a polynomial with 1 term. For example:
$3x^5$

>> A *binomial* is a polynomial with 2 terms. For example:
$3x^5 + 7x^2$

>> A *trinomial* is a polynomial with 3 terms. For example:
$3x^5 + 7x^2 - 8$

These words are handy to know when studying for the SAT. Polynomials that have more than three terms are usually called, well, just *polynomials*.

Graphing on the xy-Plane

The *xy-plane* is a mathematical construction defined by a pair of axes, where every point is labeled by a unique pair of points of the form (x, y). Figure 2-2 shows the *xy*-plane.

A wide variety of SAT questions hinge on graphs in the *xy*-plane. In this section, I refresh you on some basic graphing concepts.

Understanding the axes, the origin, and the quadrants

The *x*-axis and *y*-axis, collectively known as the *axes*, form the basis of the *xy*-plane. Essentially, these are a pair of number lines set at right angles, as shown in Figure 2-2. The arrows indicate that each axis extends infinitely in both directions. Thus, the *xy*-plane also extends infinitely in every direction.

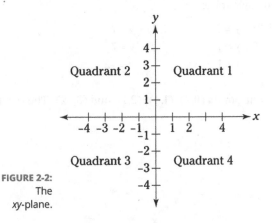

FIGURE 2-2:
The
xy-plane.

These two axes divide the *xy*-plane into four regions called *quadrants* that extend infinitely outward from the axes.

The intersection point of the two axes is called the *origin*.

Plotting coordinates on the xy-plane

Every point on the *xy*-plane is labeled uniquely as a pair of *coordinate points*, or *coordinates*, of the form (x, y). The coordinates of the origin are $(0, 0)$.

Starting from the origin, you can find the coordinates of any point by counting first on the *x*-axis, and then on the *y*-axis. This is called *plotting a point*.

Figure 2-3 shows you how to plot four points on the *xy*-plane.

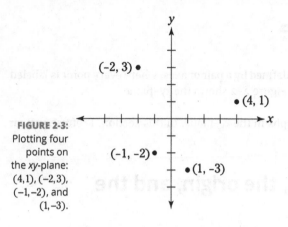

FIGURE 2-3:
Plotting four
points on
the *xy*-plane:
(4,1), (−2,3),
(−1,−2), and
(1,−3).

Plotting equations on the xy-plane

You can plot an equation on the *xy*-plane point by point, by plugging in a few values for *x* and finding the resulting value for *y* in each case. For example, to plot the equation $3x + y = 7$, plug in *x*-values of 0, 1, 2, and 3 into the equation and solve for *y*:

$$3(0) + y = 7 \qquad 3(1) + y = 7 \qquad 3(2) + y = 7 \qquad 3(3) + y = 7$$
$$0 + y = 7 \qquad 3 + y = 7 \qquad 6 + y = 7 \qquad 9 + y = 7$$
$$y = 7 \qquad\quad y = 4 \qquad\quad y = 1 \qquad\quad y = -2$$

Thus, the graph of this equation includes the four points (0,7), (1,4), (2,1), and (3,−2). The result is the line shown in Figure 2-4.

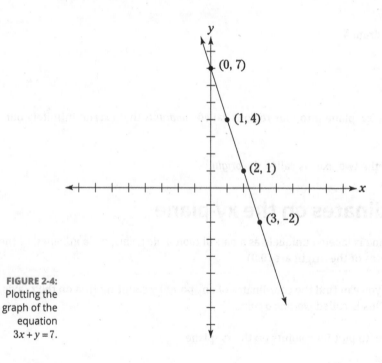

FIGURE 2-4:
Plotting the
graph of the
equation
$3x + y = 7$.

Choosing and Using a Calculator

The SAT Math Test includes two parts, one of which allows you to use a calculator. In this section, I discuss the types of calculators you can and can't use on the SAT. I also give you a list of calculator moves that every SAT student should be able to do before they start the test. Beyond this, I recommend a few more advanced graphing calculator features that you may want to play with before you take the SAT.

Knowing which calculators to avoid on the SAT

A lot of calculators are fine for the SAT, and a few just aren't. In this section, I help you rule out a few nonstarters.

No cellphone calculators allowed!

If you're like me, your main calculator is your phone. And you're probably very comfortable with it.

Unfortunately, before you start your SAT, your phone must be turned off for the duration of the test. Obviously, the SAT folks don't want you asking the internet (or your genius Uncle Harry, who works at NASA!) for help answering SAT math questions.

So for the SAT, you'll need to rely on a hand-held calculator option that *doesn't* connect to the internet. That goes for your laptop, iPad, Apple Watch, and whatever impressive technology appears next.

No fancy inputs, power cords, or paper tape

I think it's unlikely, but just in case you've got a calculator with a QWERTY keypad, you can't use that on the SAT either. Ditto for plug-in calculators or those that produce paper tape. (Do they still make those? I guess so.)

Choosing a calculator

Now that I've warned you not to use absurdly complicated calculators, I also recommend bringing one that can do more than just add, subtract, multiply, and divide. In this section, I discuss a few options.

Nice, normal calculators are fine

Casio and Texas Instruments (TI) sell a wide assortment of regular calculators and graphing calculators that you can use on the SAT. So do Radio Shack, Sharp, and a few smaller manufacturers.

A few of these have stylus inputs, which you'll have to leave untouched while you're taking the SAT. Overall, however, if you have one of these basic, garden variety calculators, you'll probably be fine.

If you've already got a calculator (other than your phone) that you're comfortable with, that's great. Or, if you have a study friend whose calculator you borrow all the time, go ahead and buy the same one.

At a minimum, your calculator should be able to crunch exponents and square roots. Calculating sines, cosines, and tangents plus their inverses also wouldn't hurt. Chances are, if your calculator can do these things, it's got enough functionality to get you through the SAT reasonably well.

Considering a graphing calculator

If you're already taking Pre-Calculus or Calculus, you may already be the proud owner and adept operator of a fancy graphing calculator. If not, you're missing out on all the fun!

But seriously, if you're shooting for an SAT math score of 700+, you should strongly consider buying a graphing calculator. Fair warning, graphing calculators are more complicated than the type that your grandmother may use for balancing her checkbook (unless, of course, your grandmother teaches physics at MIT.) But once you get the hang of it, a sophisticated calculator can really make a lot of SAT questions much simpler and way faster to answer.

Knowing some basic calculator moves

Every calculator is a little different. So if you decide to buy a new calculator especially for the SAT, be sure not to leave it in its impenetrable plastic packaging until the day of the test. Take it out and play around with it to get used to it, and, if possible, use it for a few weeks or months in your math class, as well as for taking SAT practice tests.

At a minimum, you should know how to do the following:

>> **Perform basic calculator moves:** Make sure you feel very comfortable doing basic addition, subtraction, multiplication, and division on your specific calculator.

>> **Work with decimals:** Locate the decimal point key and make sure you know how to use it.

>> **Make numbers negative:** On some calculators, the key for negating a number is distinct from the key for subtraction.

>> **Find a square root (radical):** Locate the square root key and make sure you can find square roots.

>> **Square a number:** Your calculator may have a key that looks something like x^2, used for squaring a number.

>> **Raise a number to the power of another number:** Your calculator probably has a key that looks something like x^y, which allows you to raise a number to the power of another number.

>> **Degrees versus radians:** Most calculators with trigonometric functions allow you to specify whether input values and results are to be specified in degrees or radians. Be sure you know how to make this switch so you can do it quickly on the test as needed.

Getting familiar with a few advanced calculator moves

If you've got your sights set on breaking 700 on the SAT, you should invest in a graphing calculator and spend some time figuring out how to use its more advanced features. Here are a few skills such that are likely to come in handy:

» **Fractions:** Some calculators allow you to convert decimals to fractions, perform operations on fractions, and specify that the answer be provided as a fraction. These features can be useful when a multiple-choice question provides fractional answers.

» **Parentheses:** Knowing how to group numbers using parentheses allows you to tell the calculator which operations to do first.

» **Higher-order roots (radicals):** Some calculators have a key that allows you to find higher-order radicals such as cube roots, fourth roots, and so on. But many other calculators require you to calculate higher-order roots as powers of fractions, placing the fraction in parentheses. Find out how to do this and practice it a few times so you can repeat it on the test if needed.

» **Solving equations:** The equations that you need to solve on the SAT aren't designed to be too complex, but if your calculator allows you to solve an equation for a variable, you may find this feature useful.

» **Graphing functions:** In some cases, you may find graphing a function helpful for answering a question. For example, graphing may save time solving a quadratic equation: Just graph the function and zoom in on the zeros to find x.

» **Using input-output tables:** Graphing calculators usually have a feature allowing you to make an input-output table for a function. This feature may come in handy on the SAT, so check it out.

2

Conquering the Heart of Algebra

Chapter **3**

Algebra I Reboot Part 1: Expressions

As SAT time rolls around, a lot of my students tell me that they wish they could have another go at that Algebra I class they dimly remember from 8th or 9th grade.

"Could you please explain factoring again, from the start?"

"What do you do when an equation has fractions in it?"

"Will I *really* need to know the quadratic formula?"

"How do you know when to flip the sign when you're solving an inequality?"

These are the kinds of questions my SAT math-prep students begin to ask when they feel comfortable enough to admit what they don't remember from Algebra I.

There's some good news here. If you're currently taking Algebra II or Pre-Calculus, the math that you're doing every day is a lot more difficult than what you were doing in Algebra I. So if you're hung up or unclear about some of that earlier stuff you missed, you may find this time around that it's a lot easier than you remember.

In this chapter and Chapter 4, I give you an Algebra I reboot. Here, I start with a focus on three basic algebra skills: evaluating, simplifying, and factoring expressions. First, you review how to evaluate an algebra expression by substituting (or plugging in) the values of the variables. Next, I go over how to simplify an expression by combining like terms, distributing to remove parentheses, and applying the ever-popular FOILing technique.

After that, you drill down on the most common types of factoring you're likely to see on the SAT. To begin, you review GCF factoring, which is factoring out the greatest common factor. Next, I discuss the all-important formula for difference of squares factoring, along with the lesser known but still useful formulas for factoring the sum and the difference of cubes. After that, I spend plenty of time showing you how to factor quadratic expressions. To finish up, I show you how to factor a cubic equation that fits a common pattern you might well see on the SAT.

Altogether, this chapter and Chapter 4 include almost a year's worth of Algebra I material. When you feel reasonably comfortable with this stuff, you'll be very ready to move on to the more advanced math covered in the rest of the book.

And away you go.

Evaluating Expressions

The first of three key skills for working with expressions in algebra is evaluating.

REMEMBER

Evaluating an algebra expression is simply plugging in values for the variables and then crunching the result down to a single number.

In order to evaluate an algebra expression, you have to know the value of every variable in that expression. For example,

EXAMPLE

$$3x^2 - 5xy + y$$

What is the value of this expression when $x = 2$ and $y = -7$?

To find the answer, begin by substituting 2 for x and -7 for y into the expression:

$$= 3(2)^2 - 5(2)(-7) + (-7)$$

Now, evaluate the resulting arithmetic expression using the order of operations, also known as PEMDAS (which stands for Parentheses, Exponents, Multiplication, Division, Addition, Subtraction). Begin by evaluating the exponent, then the multiplication, and finally the addition and subtraction:

$$= 3(4) - 5(2)(-7) + (-7)$$
$$= 12 + 70 - 7$$
$$= 75$$

Therefore, the correct answer is 75.

Simplifying Expressions

You can only evaluate an algebra expression when you know the values of every variable in that expression. However, even if you don't have this information, you can apply the second of the three skills for working with algebra expressions: *simplifying*.

Simplifying can tame an unfriendly-looking expression, turning it into one that you might feel more comfortable inviting over to your house for coffee and pound cake (yum!). In this section, you discover how to apply three methods for simplifying expressions: combining like terms, distributing, and FOILing.

Combining like terms

REMEMBER

Combining like terms allows you to reduce the number of terms in an algebra expression. To do this, identify any *like terms* — that is, terms that have exactly the same combination of variables — and add their coefficients.

Here's an example to make things clear.

EXAMPLE

Which of the following is equivalent to $5x^3 - 6x^2 - 9x + 10x^2 + 8x - 4$?

(A) $5x^3 + 4x^2 + x - 4$

(B) $5x^3 + 4x^2 - x - 4$

(C) $5x^3 - 4x^2 + x - 4$

(D) $5x^3 - 4x^2 - x - 4$

To begin, notice that this expression contains one pair of terms with the variable x^2 ($-6x^2 + 10x^2 = 4x^2$), and another pair of terms with the variable x ($-9x + 8x = -x$), so you can combine each of these pairs:

$$5x^3 + 4x^2 - x - 4$$

Thus, Answer B is correct.

Distributing to remove parentheses

When an expression includes parentheses, you can remove them by *distributing* the value that immediately precedes the parentheses into each term inside them. Do this according to the following rules:

REMEMBER

» When a plus sign (+) immediately precedes a set of parentheses, drop the parentheses.

» When a minus sign (–) immediately precedes a set of parentheses, flip the sign of every term inside them (from positive to negative, or from negative to positive), and then drop the parentheses.

» When a number immediately precedes a set of parentheses, multiply that number by every term inside them, and then drop the parentheses.

Here's an example to make this explanation clear:

$$6x + (2x^2 - 3) - (9x - 7) - 4(3x^2 - 5x)$$

EXAMPLE

When you simplify this expression to a polynomial, what is the coefficient of the leading term?

(A) −9

(B) −10

(C) −11

(D) −12

You can't currently combine any like terms in this expression, because no single set of parentheses contains a pair of like terms. However, you can distribute to remove parentheses.

To begin, notice that a plus sign immediately precedes the first set of parentheses, so you can just drop them:

$$= 6x + 2x^2 - 3 - (9x - 7) - 4(3x^2 - 5x)$$

The next set of parentheses is preceded by a minus sign, so distribute that minus sign to both terms inside the parentheses, flipping the sign in each case:

$$= 6x + 2x^2 - 3 - 9x + 7 - 4(3x^2 - 5x)$$

The last set of parentheses is preceded by the number -4, so to distribute this value to both terms inside, multiply each term by -4:

$$= 6x + 2x^2 - 3 - 9x + 7 - 12x^2 + 20x$$

Now, the expression includes seven terms, many of which are like terms that can be combined:

$$= -10x^2 + 17x + 4$$

As you can see, this process results in a much simpler expression whose leading coefficient is -10, so Answer B is correct.

FOILing to multiply the contents of parentheses

When a pair of parentheses in an expression are placed next to each other with no operator (such as $+$ or $-$) between them, the expressions inside the parentheses can be multiplied using distribution. In this case, every term inside one set of parentheses must be multiplied by every term inside the other set.

In this section, I show you how to do classic FOILing to multiply a pair of binomials (two-term polynomials). Then I show you an organized way to do the same process when one or both polynomials have more than two terms.

FOILing a pair of binomials

Most math teachers use the acronym FOIL to describe this process. FOIL stands for First, Outside, Inside, Last. Here's an example that describes the process:

EXAMPLE

If $(3x + 4)(5x - 2) = ax^2 + bx + c$ for all values of x, then what does $a + b + c$ equal?

To find this answer, you need to multiply four times:

Multiply the two *first* terms: $3x$ times $5x$ equals $15x^2$.

Multiply the two *outside* terms: $3x$ times -2 equals $-6x$.

Multiply the two *inside* terms: 4 times $5x$ equals $20x$.

Multiply the two *last* terms: 4 times -2 equals -8.

The result is the following four-term expression:

$$= 15x^2 - 6x + 20x - 8$$

You can simplify this expression further by combining like terms:

$$= 15x^2 + 14x - 8$$

Thus, $a + b + c = 15 + 14 - 8$, so the correct answer is 21.

FOILing with larger polynomials

When one or more of the polynomials you're multiplying has more than two terms, I recommend taking an organized approach.

EXAMPLE

$$(x+3)(4x^2 - 5x + 1)$$

Which of the following is an equivalent form of this expression?

(A) $4x^3 + 7x^2 + 14x + 3$

(B) $4x^3 + 7x^2 - 14x + 3$

(C) $4x^3 - 7x^2 + 14x + 3$

(D) $4x^3 - 7x^2 - 14x + 3$

To multiply any pair of polynomials, you need to multiply each term in the first polynomial by every term in the second, and then combine like terms to simplify the result. To begin, multiply x by $4x^2 - 5x + 1$:

$$4x^3 - 5x^2 + x$$

Next, multiply 3 by $4x^2 - 5x + 1$, placing each like term below the corresponding like term from the first step:

$$4x^3 - 5x^2 \quad + x$$
$$12x^2 - 15x + 3$$

Doing the multiplication in this way makes errors less likely, and makes combining like terms much easier. Just add each column, as follows:

$$= 4x^3 + 7x^2 - 14x + 3$$

Therefore, Answer B is correct.

Factoring Expressions

The third skill for working with expressions is *factoring* — changing an expression to two or more smaller expressions that are multiplied together.

In a sense, factoring is the flipside of simplifying. When you simplify an expression, you often *remove* parentheses; in contrast, factoring an expression *introduces* one or more sets of parentheses to the expression.

In this section, you discover five common ways to factor expressions.

GCF Factoring

GCF factoring is the most basic type of factoring: pulling the greatest common factor (GCF) out of each term of the expression. Here is an example:

$$20x^5 + 12x^3 - 16x^2$$

Which of the following is the completely factored form of this expression?

(A) $2x^2(5x^3 + 3x - 4)$

(B) $2x^2(10x^3 + 6x - 8)$

(C) $4x^2(5x^3 + 3x - 4)$

(D) $4x^2(10x^3 + 6x - 8)$

This expression includes three terms whose coefficients (20, 12, and –16) are all divisible by 4. Additionally, each term has a variable of x raised to a different power, the lowest of which is 2. Thus, the GCF of this expression is $4x^2$. To factor out this value from each term in the equation, write the value $4x^2$ outside a pair of parentheses, and then divide each term in the equation by this value:

$$= 4x^2(5x^3 + 3x - 4)$$

Thus, Answer C is correct.

$$45x^7y^2 - 27x^5y^3 + 36x^3y^4$$

Which of the following is the greatest common factor of this expression?

(A) $9x^3y^2$

(B) $9x^3y^4$

(C) $9x^7y^2$

(D) $15x^7y^4$

This expression includes three terms whose coefficients (45, –27, and 36) are all divisible by 9, which rules out Answer D. Additionally, each term has a variable of x raised to a different power, the lowest of which is 3, which rules out Answer C. And each term has a variable of y raised to a different power, the lowest of which is 2, which rules out Answer B, so Answer A is correct.

Difference of squares

The formula for factoring a difference of squares is

$$x^2 - y^2 = (x + y)(x - y)$$

Step back from this formula for a moment, and notice the following:

>> The left side of the formula includes exactly two terms.

>> Both terms are perfect squares.

>> Both terms are separated by a minus sign (–).

Whenever an expression includes these features, you can use this formula to factor it.

Table 3-1 includes a variety of examples that show you how to factor using the difference of squares formula. Here's an SAT example:

TABLE 3-1 **Difference of Squares Factoring**

Before Factoring	After Factoring
$x^2 - 25$	$(x+5)(x-5)$
$16x^2 - 1$	$(4x+1)(4x-1)$
$36x^2 - 49y^2$	$(6x+7y)(6x-7y)$
$x^6 - y^{10}$	$(x^3 + y^5)(x^3 - y^5)$
$81x^8 - 100y^{14}$	$(9x^4 + 10y^7)(9x^4 - 10y^7)$
$\sin^2 x - \cos^2 x$	$(\sin x + \cos x)(\sin x - \cos x)$
$53^2 - 47^2$	$(53 + 47)(53 - 47)$

EXAMPLE

$$4x^2 - 9$$

This expression is equivalent to which of the following?

(A) $(2x-3)(2x-3)$

(B) $(2x+3)(2x-3)$

(C) $(x-3)(4x-3)$

(D) $(x+3)(4x-3)$

Note that GCF factoring is not an option, because the greatest common factor of $4x^2$ and -9 is 1. However, the expression is the difference of two squares, so you can factor it as follows:

$$= (2x+3)(2x-3)$$

Therefore, Answer B is correct.

This method works equally well with arithmetic expressions. Here is an example:

EXAMPLE

Which of the following is equivalent in value to $987^2 - 123^2$?

(A) $(987+123)^2$

(B) $(987-123)^2$

(C) $(987+123)(987-123)$

(D) $(987+123)^2 - 2(987-123)^2$

Use the difference of squares formula to factor as follows:

$$987^2 - 123^2 = (987+123)(987-123)$$

So Answer C is correct.

Here's an example that uses both GCF and difference of squares factoring:

EXAMPLE

$$\frac{4x^2-9}{4x+6}$$

Which of the following is equivalent to this expression?

(A) $\frac{x+3}{2}$

(B) $\frac{x-3}{2}$

(C) $\frac{2x+3}{2}$

(D) $\frac{2x-3}{2}$

To begin, use difference of squares factoring in the numerator, and factor out the GCF in the denominator:

$$\frac{4x^2-9}{4x+6}=\frac{(2x+3)(2x-3)}{2(2x+3)}$$

Now, cancel factors of $2x+3$ in both the numerator and denominator:

$$=\frac{2x-3}{2}$$

Therefore, Answer D is correct.

You can use the principle behind difference of squares factoring even when a polynomial isn't factorable into two polynomials. Here is an example:

EXAMPLE

Which of the following expressions is equivalent to $x-5$?

(A) $(\sqrt{x}+\sqrt{5})^2$

(B) $(\sqrt{x}-\sqrt{5})^2$

(C) $(\sqrt{x}+\sqrt{5})(\sqrt{x}-\sqrt{5})$

(D) None of these expressions is equivalent to $x-5$.

Although $x-5$ isn't factorable, if you follow the rules of difference of squares factoring, you find the following surprising result:

$$x-5=(\sqrt{x}+\sqrt{5})(\sqrt{x}-\sqrt{5})$$

To verify this equivalence, FOIL the expression $(\sqrt{x}+\sqrt{5})(\sqrt{x}-\sqrt{5})$:

$$(\sqrt{x}+\sqrt{5})(\sqrt{x}-\sqrt{5})=x-\sqrt{5}\sqrt{x}+\sqrt{5}\sqrt{x}-5=x-5$$

Therefore, Answer C is correct.

A particularly tricky question of this sort that you may find on the SAT provides you with the sum of squares and asks you to find an equivalent form of the expression. Here is an example:

EXAMPLE

Which of the following expressions is equivalent to $4x+9$?

(A) $(\sqrt{2x}+3)(\sqrt{2x}-3)$

(B) $(\sqrt{2x}+3i)(\sqrt{2x}-3i)$

(C) $(2\sqrt{x}+3)(2\sqrt{x}-3)$

(D) $(2\sqrt{x}+3i)(2\sqrt{x}-3i)$

To be clear, the polynomial $4x+9$ is not factorable into a pair of polynomials. But the question is asking only for an equivalent form of the original polynomial. To begin, express this polynomial as the difference of two terms:

$$4x+9 = 4x-(-9)$$

Now, simplify $\sqrt{4x}$ and $\sqrt{-9}$:

$$\sqrt{4x} = 2\sqrt{x} \qquad \sqrt{-9} = 3i$$

Thus, using the formula for the difference of squares, you can express $4x+9$ equivalently as follows:

$$(2\sqrt{x}+3i)(2\sqrt{x}-3i)$$

Therefore, Answer D is correct.

Keep an eye out for SAT questions that provide an opportunity to use this trick, especially if you're aiming to get 700 or better on the math section!

Sum and difference of cubes

The formulas for factoring a sum or a difference of cubes are not used nearly as much on the SAT as the one for factoring the difference of squares (which I describe in the previous section). Even so, they may come in handy, so if you're shooting for a score of 700 or better, take a moment to recall these two formulas:

$$x^3+y^3 = (x+y)(x^2-xy+y^2)$$
$$x^3-y^3 = (x-y)(x^2+xy+y^2)$$

Notice that the two formulas differ only in the placement of the minus sign on the right side of the equals sign. Here's an SAT question that draws upon this information:

$$27x^3+125y^3$$

Which of the following is equivalent to this expression?

(A) $(3x+5y)(9x^2-15xy+25y^2)$

(B) $(3x-5y)(9x^2+15xy+25y^2)$

(C) $(9x+25y)(3x^2-15xy+25y^2)$

(D) $(9x-25y)(3x^2+15xy+5y^2)$

The question asks you to find the sum of cubes, so if you know the pattern of minus signs for this formula, you can rule out Answers B and D quickly. Applying the formula more closely provides the following result:

$$27x^3+125y^3 = (3x+5y)(9x^2-15xy+25y^2)$$

Therefore, Answer A is correct.

Factoring quadratic trinomials

A *quadratic trinomial* (polynomial with three terms) is an expression that has the form $ax^2 + bx + c$, where a, b, and c are all real numbers. When a, b, and c are all integers, in many cases (especially on the SAT!), the resulting quadratic expression can be factored into a pair of binomials.

In this section, I first show you how to factor a quadratic expression of the form $x^2 + bx + c$, that is, when the leading coefficient $a = 1$. Then you work through the slightly more complicated case where $a \neq 1$.

Easy quadratic factoring: When the leading coefficient is 1

To factor a quadratic trinomial of the form $x^2 + bx + c$ (that is, with a leading coefficient of 1), focus on the b and c values. If the expression is factorable, you will be able to find a pair of integers that do the following:

» Add up to b

» Multiply to c

Here is an example:

EXAMPLE

Factor the following trinomial:

$$x^2 + 8x + 15$$

In this case, the two integers you're looking for are 3 and 5, because $3 + 5 = 8$ and $3 \times 5 = 15$. So you can factor this trinomial as follows:

$$= (x + 3)(x + 5)$$

EXAMPLE

When the b value is negative and the c value is positive, the two numbers you're looking for are both negative. To illustrate this point, I'll make only a small change to the previous example. Suppose you want to factor this polynomial:

$$x^2 - 8x + 15$$

This time, the b value -8 is negative and the c value 15 remains positive. Thus, you're looking for a pair of negative numbers that add up to -8 and multiply to 15. This time, the pair of numbers -3 and -5 fits the bill, so you can factor as follows:

$$= (x - 3)(x - 5)$$

When the b value is negative and the c value is positive, the pair of numbers you're looking for include a positive number and a negative number. To see how this works, consider this example:

$$x^2 + 6x - 27$$

In this case, the two values 9 and -3 work, because $9 + (-3) = 6$ and $9 \times (-3) = -27$, so you can factor in this way:

$$= (x + 9)(x - 3)$$

When the b and c values are both negative, the pair of numbers you're looking for include a positive number and a negative number, with a slight difference. To make sure you understand this fine point, I'm making only a slight change to the previous example:

$$x^2 - 6x - 27$$

In this case, the two values 3 and -9 work, because $3+(-9)=-6$ and $9\times(-3)=-27$, so you can factor in this way:

$$= (x+3)(x-9)$$

Notice that in this case, the two signs in the result are flipped.

Table 3-2 shows four examples that are similar to these. It also gives you a few more examples to think about.

TABLE 3-2   **Difference of Squares Factoring**

Before Factoring	After Factoring
$x^2+7x+10$	$(x+2)(x+5)$
$x^2-7x+10$	$(x-2)(x-5)$
x^2+5x-6	$(x+6)(x-1)$
x^2-5x-6	$(x+1)(x-6)$
$x^2+8xy+12y^2$	$(x+2y)(x+6y)$
x^4+6x^2+8	$(x^2+2)(x^2+4)$
$\sin^2 x+10\sin x+9$	$(\sin x+9)(\sin x+1)$

Difficult quadratic factoring: When the leading coefficient isn't 1

On the SAT, you may not run into a question that includes a quadratic trinomial with a leading coefficient of 1. However, because the test tends to incorporate new material over time, you might do well to keep this information in your back pocket.

On the bright side, in some cases, you may be able to factor this type of polynomial by using GCF factoring first. Here is an example:

EXAMPLE

$$6x^2-12x-18$$

This expression looks like trouble because the numbers are large, but if you begin by factoring out the GCF, you get this result:

$$= 6(x^2-2x-3)$$

Now, the trinomial inside the parentheses has a leading coefficient of 1, so you can factor it easily using the method I discuss in the previous section:

$$= 6(x+1)(x-3)$$

TIP

Keep a sharp lookout on the SAT for opportunities like this example to let GCF factoring make your life easier!

However, in some cases, GCF factoring just isn't an option. In this case, assuming the trinomial is factorable, you can use *factoring by grouping*. Here is an example:

EXAMPLE

$$6x^2 - 7x - 5$$

To factor this trinomial, you're going to need a slightly more complicated method. Fortunately, if you already know how to do GCF factoring and how to factor a quadratic trinomial with a leading coefficient of 1, you're already set up for success.

1. **Find two integers that both add up to b and multiply to ac (that is, $a \times c$).** In this case, $b = -7$ and $ac = -30$, so the two integers you're looking for are 3 and –10.

2. **Split the x term of the trinomial into a pair of terms whose coefficients are the integers you found in Step 1.**

   $$6x^2 - 7x - 5$$
   $$= 6x^2 + 3x - 10x - 5$$

   As you can see, this step does not change the value of the expression that you started with.

3. **Group this resulting polynomial into two pairs of binomials separated by parentheses and a plus sign.**

   $$= (6x^2 + 3x) + (-10x - 5)$$

   Again, this step preserves the expression.

4. **Factor out the GCF of both groups (making sure that the same binomial remains inside the parentheses).** In this case, you can factor out $3x$ and –5:

   $$= 3x(2x + 1) - 5(2x + 1)$$

5. **Add the two outside numbers as one factor, then copy the contents of the parentheses as the other factor.** This step is easier to do than to explain:

   $$= (3x - 5)(2x + 1)$$

The result is the factored version of your original trinomial. You can verify this by FOILing it:

$$= 6x^2 + 3x - 10x - 5$$
$$= 6x^2 - 7x - 5$$

Notice here that the intermediate step in the FOILing process reveals the split that you made to the polynomial in Step 2.

Factoring cubic expressions

In some cases, you can factor a cubic expression by applying GCF factoring and then quadratic factoring. This is often possible when a cubic expression has only three terms, with no constant term. Here is an example:

EXAMPLE

Factor the following expression:

$$3x^3 - 27x^2 + 42x$$

Begin by factoring out $3x$ from every term:

$$= 3x(x^2 - 9x + 14)$$

Now, you can factor the trinomial that remains inside the parentheses using the easier quadratic method:

$$= 3x(x - 2)(x - 7)$$

When a cubic expression has four terms, however, it may be factorable by a different method called *grouping*. This is similar to the method for factoring difficult quadratic trinomials described earlier in this chapter.

For example, suppose you want to factor this cubic polynomial:

$$5x^3 - 7x^2 + 15x - 21$$

Before you begin, you may notice that the coefficients of the first pair of terms in this expression (5 and −7) and the second pair of terms (15 and −21) are directly proportional — in this case, by a factor of 3. When this type of proportionality holds for a cubic equation, you can use grouping to factor:

1. **Group the polynomial into two pairs of binomials separated by parentheses and a plus sign.**

$$= (5x^3 - 7x^2) + (15x - 21)$$

   This step should remind you of Step 3 in the previous section. The next two steps here mimic Steps 4 and 5 in that section.

2. **Factor out the GCF of both groups (making sure that the same binomial remains inside the parentheses).** Here, you can factor out x^2 and 3:

$$= x^2(5x - 7) + 3(5x - 7)$$

3. **Add the two outside numbers as one factor, then copy the contents of the parentheses as the other factor.** This step is still easier to do than to explain:

$$= (x^2 + 3)(5x - 7)$$

In some cases, you may be able to apply GCF factoring before you factor by grouping, and then tweak your result further using difference of squares factoring. This example puts it all together:

$$60x^4 - 20x^3 - 15x^2 + 5x$$

To begin, notice that every term here has a factor of $5x$:

$$= 5x(12x^3 - 4x^2 - 3x + 1)$$

WARNING

This factor carries along to the end of the problem, so be sure not to misplace it along the way! What's left inside the brackets is a garden-variety factorable cubic equation with proportional coefficients, as I explain previously, so you can perform Steps 3, 4, and 5 as follows:

$$= 5x[4x^2(3x - 1) - 1(3x - 1)]$$
$$= 5x[4x^2(3x - 1) - 1(3x - 1)]$$
$$= 5x(4x^2 - 1)(3x - 1)$$

This result may look complete, but notice that the second factor, $4x^2 - 1$, is the difference of squares, so you can factor it further:

$$= 5x(2x + 1)(2x - 1)(3x - 1)$$

In all likelihood, you probably won't have to factor anything nearly this complex on the SAT. But knowing all the factoring methods I discuss throughout this chapter will definitely help you achieve a higher score.

IN THIS CHAPTER

» Solving basic algebra equations by balancing the scale and isolating the variable

» Working with equations that have more than one variable

» Solving two-term quadratic equations quickly

» Using factoring and the quadratic formula to solve quadratic equations

» Simplifying rational equations to solve them

» Handling inequalities by flipping the sign

Chapter **4**

Algebra I Reboot Part 2: Equations and Inequalities

n Chapter 3, you discover (or rediscover) a ton of Algebra I skills you may have missed the first time around. Here, you apply these moves to solve a wide variety of algebra equations similar to those you're sure to see when you take the SAT Math Test.

To begin, you start with a review of how to use algebra to solve relatively simple equations. Next, you work with equations that have more than one variable. After that, you face down more difficult equations, such as quadratic and rational equations. Finally, to round things out, you solve inequalities by knowing when and how to flip the inequality sign.

This chapter, together with Chapter 3, covers an entire year's worth of algebra, so feel free to take your time with it. When you understand this stuff fully, you'll find that a lot of tough-looking SAT math questions suddenly seem a lot more manageable.

Solving Simple and Intermediate Algebra Equations

Solving equations is the main event in algebra. In this section, you use the tools given in Chapter 3 to solve a variety of increasingly complex algebra equations.

Solving basic algebra equations

You can solve some algebra equations just by inspection — that is, just by looking at them and thinking about what the variable must be. For example, the following four algebra equations are progressively more difficult. How many can you solve without writing anything down?

$$x + 7 = 10 \qquad 3x + 2 = 8 \qquad 8x = 144 - 4x \qquad 9x + 26 = -5x - 16$$

If you struggled and finally gave up on at least one of these equations, you can probably see the need for a more systematic method for solving equations that arise when building 60-story towers, mapping the human genome, or sending rockets to Saturn.

You start by solving basic equations by *isolating the variable*, working step by step to discover the value of the variable by getting it alone on one side of the equals sign. When the variable is left standing by itself with no place to hide, whatever's left on the other side of the equals sign is the solution.

Each step along the way, you use the *balance scale method*, which encourages you to think of an equation as a scale with two perfectly balanced sides: If you remove the same amount of weight from both sides of the scale, it stays in balance. In an equation, when you perform the same operation on both sides, they both remain equal. Here is an example:

EXAMPLE

If $8x = 144 - 4x$, then what does $2x - 7$ equal?

(A) 12

(B) 14

(C) 15

(D) 17

This is a common type of SAT question that asks you to solve one equation, and then plug that answer into another expression and evaluate the result. Begin by solving the equation. To make it simpler, add $4x$ to both sides:

$$12x = 144$$

At this point, you may already see the answer, but if you have any doubt, divide both sides by 12:

$$x = 12$$

WARNING

You're not done! The question isn't asking for the value of x, but rather the value of $2x - 7$:

$$2x - 7 = 2(12) - 7 = 24 - 7 - 17$$

Therefore, Answer D is correct.

As another example, consider the following question, which includes an equation from earlier in this section:

EXAMPLE

$$9x + 26 = -5x - 16$$

What is the value of |x| in this equation?

Again, you want to solve the equation and then plug the result into an expression. This time, begin by adding 5x to both sides:

$$14x + 26 = -16$$

Now subtract 26 from both sides:

$$14x = -42$$

To finish up, divide by 14:

$$x = -3$$

TIP

You're not done! This question doesn't give you four answers labeled A through D, so it's an open-ended (student-produced response) question. Remember that you can't place a negative number into the grid. Fortunately, the question asks not for x but rather |x|:

$$|x| = |-3| = 3$$

Therefore, the correct answer is 3.

More complicated equations require you to distribute in order to remove parentheses on one or both sides of the equation before you can combine like terms and solve. Here is an example:

EXAMPLE

$$4x + (7 - 6x) = 19 - 4(2x - 9)$$

What is the value of x in this equation?

To solve this equation, begin by dropping the parentheses on the left side of the equation and distributing on the right side:

$$4x + 7 - 6x = 19 - 8x + 36$$

Next, simplify both sides of the equation:

$$7 - 2x = 55 - 8x$$

Now, solve the equation using methods from earlier in this chapter that you already know and love:

$$7 + 6x = 55$$
$$6x = 48$$
$$x = 8$$

Therefore, the correct answer is 8.

Working with equations that have more than one variable

In most cases, you can't solve an equation that has more than one variable, because it produces infinitely many solutions. But SAT questions like these skirt around this issue in a variety of ways. In this section, I give you a few practical pointers for answering these types of questions.

Isolating a specific variable

One common type of SAT math question presents you with an equation in two or more variables and asks you to solve it for a given variable. To do this, isolate that variable on one side of the equation and get all the other variables over to the other side. Here is an example:

EXAMPLE

If $x = \dfrac{2a + 3b^2}{c}$, what is the value of a in terms of b, c, and x?

(A) $\dfrac{3b + 2c^2}{x}$

(B) $\dfrac{cx + 2b^2}{3}$

(C) $\dfrac{bx - 2c^2}{3}$

(D) $\dfrac{cx - 3b^2}{2}$

To solve, isolate the variable a. Begin by multiplying both sides of the equation $x = \dfrac{2a + 3b^2}{c}$ by c:

$$cx = 2a + 3b^2$$

Next, subtract $3b^2$ from both sides:

$$cx - 3b^2 = 2a$$

To finish, divide both sides by 2:

$$\frac{cx - 3b^2}{2} = a$$

Therefore, Answer D is correct.

A common type of SAT math question gives you a geometric formula and asks you to solve it for a specific variable. Here is an example:

EXAMPLE

$$A = \frac{b_1 + b_2}{2}h$$

This formula is for the area A of a trapezoid based on its height h and the lengths of its two bases, b_1 and b_2. Which of the following allows you to find the length of base b_2 based on the values of the other three variables?

(A) $b_2 = \dfrac{A - b_1}{2h}$

(B) $b_2 = \dfrac{2A + b_1}{h}$

(C) $b_2 = \dfrac{A}{2h} + b_1$

(D) $b_2 = \dfrac{2A}{h} - b_1$

To solve for b_2, begin by dividing both sides of the equation by h:

$$\frac{A}{h} = \frac{b_1 + b_2}{2}$$

Next, multiply both sides by 2:

$$\frac{2A}{h} = b_1 + b_2$$

Finally, subtract b_1 from both sides:

$$\frac{2A}{h} - b_1 = b_2$$

Therefore, Answer D is correct.

Factoring to solve for a variable

In some cases, when solving an equation for a variable, you may need to factor to isolate that variable. Here is an example:

EXAMPLE

If $2xy + 5 = -3y + 12x$, what is the value of y in terms of x?

(A) $\dfrac{12x + 5}{2x + 3}$

(B) $\dfrac{12x - 5}{2x + 3}$

(C) $\dfrac{12x + 5}{2x - 3}$

(D) $\dfrac{12x - 5}{2x - 3}$

To begin, isolate the terms that include y on one side of the equals sign:

$$2xy + 3y = 12x - 5$$

Now, factor out y on the left side of the equation:

$$y(2x + 3) = 12x - 5$$

To finish up, divide out $2x + 3$ on both sides. I do this in two steps, so you can see how it works:

$$\frac{y(2x + 3)}{2x + 3} = \frac{12x - 5}{2x + 3}$$
$$y = \frac{12x - 5}{2x + 3}$$

Therefore, Answer B is correct.

Solving for an expression that contains more than one variable

In some cases, an SAT question will give you an equation that has more than one variable and ask you to find the value of an expression that includes all the variables. In these cases, you need to think of the two variables as a single unit and figure out how to produce the expression that the question provides, as in this example:

EXAMPLE

If $a + b + 5 = 51$, what is the value of $2(a + b) - 3$?

Your obvious first step is to subtract 5 from both sides:

$$a + b = 46$$

Now, you may feel stuck, because you can't solve for either a or b. However, step back from the equation and notice that the question is asking you for the value of $2(a + b) - 3$. Thus, you can substitute 46 for $a + b$ and evaluate:

$$2(46) - 3 = 92 - 3 = 89$$

Therefore, the correct answer is 89.

Here's another example: If $4m - 6n + 17 = 29$, what does $2m - 3n + 10$ equal?

This time, begin by subtracting 17 from both sides:

$$4m - 6n = 12$$

As with the previous example, you can't solve this equation for either variable. The insight here is to see that you can build $2m - 3n$ by dividing each side of the equation by 2:

$$2m - 3n = 6$$

Now, just add 10 to both sides of the equation and simplify:

$$2m - 3n + 10 = 6 + 10 = 16$$

Therefore, the correct answer is 16.

Keep an eye open for opportunities to use these tricks to answer SAT math questions!

Solving More Difficult Equations

In this section, I turn up the heat a bit, showing you how to answer a variety of common SAT math questions that require more difficult solving techniques.

Solving quadratic equations

Quadratic equations can get tricky because, depending on the equation, there are a variety of ways to solve them. In this section, I start off with simpler two-term quadratic equations, which you can solve by isolating the variable, as I discuss in Chapter 3. Next, I show you how to solve trickier three-term quadratic equations.

Solving two-term quadratic equations

In some cases, quadratic equations include only two terms. These tend to be the simplest quadratic equations to solve. In this section, I show you how to solve the two types of two-term quadratic equations.

SOLVING $ax^2 + c = 0$

You can solve a quadratic of the form $ax^2 + c = 0$ by isolating the variable x. Here is an example:

EXAMPLE

If $3x^2 - 4 = 0$, then one possible value of x is

(A) $\frac{\sqrt{2}}{3}$

(B) $\frac{\sqrt{3}}{3}$

(C) $\frac{2\sqrt{2}}{3}$

(D) $\frac{2\sqrt{3}}{3}$

To solve, add 4 to both sides, and then divide by 3:

$$3x^2 = 4$$

$$x^2 = \frac{4}{3}$$

Now, take the square root of both sides:

$$x = \pm\frac{2}{\sqrt{3}}$$

To make this expression look more like the answers, rationalize the denominator by multiplying both the numerator and denominator by $\sqrt{3}$ and then simplifying:

$$= \pm \frac{2}{\sqrt{3}} \frac{\sqrt{3}}{\sqrt{3}} = \pm \frac{2\sqrt{3}}{3}$$

Thus, Answer D is correct.

SOLVING $ax^2 + bx = 0$

Solving a two-term quadratic equation of the form $ax^2 + bx = 0$ is also fairly straightforward. To do this, begin by factoring out an x, and then split the equation into two equations, solving each separately. The following example shows you how:

EXAMPLE

If $-5x^2 + 4x = 0$, then what is the non-zero solution to this equation?

To solve a quadratic equation of this type, begin by factoring out an x:

$$x(-5x + 4) = 0$$

Now, you have two factors that, when multiplied, equal 0. Therefore, one of these two factors equals 0, so you can split the equation into two and solve each one separately:

$$
\begin{array}{ccc}
 & & -5x + 4 = 0 \\
x = 0 & \text{OR} & -5x = -4 \\
 & & x = \frac{4}{5}
\end{array}
$$

Therefore, the correct answer is $\frac{4}{5}$ or 0.8. Either answer is correct, but when you grid in the decimal value, write it as .8 *without* the leading zero.

Solving three-term quadratic equations

A quadratic equation that includes all three possible terms is of the form

$$0 = ax^2 + bx + c$$

FACTORING TO SOLVE QUADRATICS

In Chapter 3, I show you how to factor quadratic expressions. This skill comes in handy when solving quadratic equations, some of which can be solved by factoring. Here is an example:

EXAMPLE

$$x^2 - 6x = 40$$

If n is a solution to this quadratic equation and $n > 0$, then what is the value of n?

To begin, subtract 40 from each side, turning the equation into a quadratic expression set to equal 0:

$$x^2 - 6x - 40 = 0$$

Next, factor this expression as I show you in the previous section:

$$(x + 4)(x - 10) = 0$$

Because this expression is now written as a pair of factors that equal 0, one of these two factors must equal 0, so:

$$x + 4 = 0 \qquad \text{OR} \qquad x - 10 = 0$$

Thus, $x = -4$ and $x = 10$ are the two solutions to the equation. And $n > 0$, so the correct answer is 10.

Here's another example that requires you to factor a more difficult expression:

EXAMPLE

$$-4x^2 = -15x + 9$$

If u and v are possible solutions to this equation, then what does $|u - v|$ equal?

As in the previous example, begin by bringing all three terms to the same side of the equation. When you do this, I recommend that you move the leading term $-4x^2$ across the equals sign so that it becomes positive:

$$0 = 4x^2 - 15x + 9$$

Now, factor the right side of this equation as I show you in Chapter 3:

$$0 = 4x^2 - 12x - 3x + 9$$
$$0 = 4x(x - 3) - 3(x - 3)$$
$$0 = (4x - 3)(x - 3)$$

Thus, you can break the equation into two separate equations and solve each one separately:

$$0 = 4x - 3$$
$$3 = 4x \qquad \text{OR} \qquad \begin{array}{l} 0 = x - 3 \\ 3 = x \end{array}$$
$$\frac{3}{4} = x$$

Thus, u and v are these two solutions in some order, so

$$|u - v| = \left| 3 - \frac{3}{4} \right| = \frac{9}{4}$$

On the SAT, you can grid-in this answer either as the improper fraction $\frac{9}{4}$ or in decimal form as 2.25.

WARNING

When gridding in this answer as a fraction, *don't* enter it as the mixed number 2¼, because this will show up as 21/4, which is incorrect. (For more on entering answers to grid-in questions, see Chapter 1.)

USING THE QUADRATIC FORMULA

The quadratic formula is a staple of every algebra class. It allows you to solve any quadratic equation of the form $ax^2 + bx + c = 0$ for x in terms of a, b, and c:

$$x = \frac{-b \pm \sqrt{b^2 - 4ac}}{2a}$$

You've probably already seen this famous (or infamous) formula before, but in any case, you should memorize it for use on the SAT.

The quadratic formula is most useful for solving quadratic equations that don't solve nicely by factoring them. Here is an example:

$$8x^2 - 20x - 16 = 0$$

Which of the following gives both solutions to this equation?

(A) $\dfrac{5 \pm \sqrt{7}}{4}$

(B) $\dfrac{-5 \pm \sqrt{7}}{4}$

(C) $\dfrac{5 \pm \sqrt{57}}{4}$

(D) $-\dfrac{5 \pm \sqrt{57}}{4}$

Begin by dividing both sides of the equation by 4:

$$2x^2 - 5x - 4 = 0$$

Unfortunately, you can't factor the trinomial $2x^2 - 5x - 4$ into a pair of binomials with integer coefficients. So, if you can't use your calculator — that is, if you're taking the No Calculator section of the SAT! — your only alternative is the quadratic formula. Begin by plugging in 2 for a, -5 for b, and -4 for c:

$$x = \frac{-b \pm \sqrt{b^2 - 4ac}}{2a} = \frac{-(-5) \pm \sqrt{(-5)^2 - 4(2)(-4)}}{2(2)}$$

Simplify as much as possible:

$$= \frac{5 \pm \sqrt{25 + 32}}{4} = \frac{5 \pm \sqrt{57}}{4}$$

Therefore, Answer C is correct.

Sometimes when you use the quadratic formula, you may need to tweak your result by simplifying the radical or reducing a rational expression. Here is an example:

$$x^2 + 12x - 7 = 0$$

Which of the following is the solution set for this quadratic equation?

(A) $6 \pm \sqrt{29}$

(B) $-6 \pm \sqrt{29}$

(C) $6 \pm \sqrt{43}$

(D) $-6 \pm \sqrt{43}$

Begin by plugging in 1 for a, 12 for b, and -7 for c into the quadratic equation:

$$x = \frac{-b \pm \sqrt{b^2 - 4ac}}{2a}$$

$$= \frac{-12 \pm \sqrt{12^2 - 4(1)(-7)}}{2(1)}$$

$$= \frac{-12 \pm \sqrt{144 + 28}}{2}$$

$$= \frac{-12 \pm \sqrt{172}}{2}$$

This result still doesn't look like any of the four answers, but it can be simplified by splitting the radical into two factors:

$$= \frac{-12 \pm \sqrt{4}\sqrt{43}}{2}$$

$$= \frac{-12 \pm 2\sqrt{43}}{2}$$

Now, you can reduce the numerator and denominator by a factor of 2 as follows:

$$= \frac{2(-6 \pm \sqrt{43})}{2}$$

$$= -6 \pm \sqrt{43}$$

Therefore, Answer D is correct.

Solving rational equations

A *rational equation* includes at least one term that has an element with a variable in either the numerator or denominator of a fraction.

In an Algebra II or Pre-Calculus class, you may go very deeply into rational equations and functions. On the SAT, however, you need to know only a handful of skills for solving rational equations. In this section, I go over these skills.

Multiplying by the denominator

When one side of an equation is a single rational term, multiply both sides by the denominator of this term. Here is an example:

$$\frac{3x}{x+1} = -6$$

EXAMPLE

For this equation, what is the value of $|x|$?

To answer this question, multiply both sides by the denominator, $x+1$, and simplify:

$$\frac{3x}{x+1}(x+1) = -6(x+1)$$

$$3x = -6x - 6$$

Multiplying by the denominator cancels it out and makes the equation non-rational, so you can solve it using methods you already know:

$$9x = -6$$

$$x = -\frac{6}{9} = -\frac{2}{3}$$

Therefore, $|x| = \frac{2}{3}$, which you can also enter into the grid as either .666 or .667.

Cross-multiplying

When both sides of a rational equation are rational expressions, cross-multiplication is the best tool for the job. To cross-multiply, follow these steps:

1. Multiply the numerator of the left side by the denominator of the right side.

2. Multiply the numerator of the right side by the denominator of the left side.

3. Set these two values equal to each other.

EXAMPLE

Here's an example that gives you practice with cross-multiplication:

$$\frac{x+2}{x-3} = \frac{x+4}{x-2}$$

What is the value of x in this equation?

To solve, cross-multiply:

$$(x+2)(x-2) = (x+4)(x-3)$$

The result is an equation that no longer includes rational expressions. To complete the problem, FOIL and then isolate x:

$$x^2 - 4 = x^2 + x - 12$$
$$-4 = x - 12$$
$$8 = x$$

Getting a common denominator

When a rational equation includes rational terms mixed in with other terms, finding a common denominator is usually the way to solve it. Here is an example:

EXAMPLE

$$\frac{k}{5} = \frac{k+4}{3} - 2k$$

What value of k solves this equation?

This equation includes two rational terms ($\frac{k}{5}$ and $\frac{k+4}{3}$) and a non-rational term ($-2k$). The common denominator for the two rational terms is 15, so you want to rewrite the equation as three terms with denominators of 15. Do the following:

» Multiply the terms of $\frac{k}{5}$ by a factor of 3

» Multiply the terms of $\frac{k+4}{3}$ by a factor of 5, and

» Multiply the terms of $-2k$ by a factor of 15

I do this in two steps:

$$\frac{k}{5} \times \frac{3}{3} = \frac{k+4}{3} \times \frac{5}{5} - 2k \times \frac{15}{15}$$
$$\frac{3k}{15} = \frac{5(k+4)}{15} - \frac{30k}{15}$$

When every term in an equation has the same denominator, you can multiply the entire equation by that denominator, canceling it out:

$$3k = 5(k+4) - 30k$$

Now, distribute, simplify, and solve the equation:

$$3k = 5k + 20 - 30k$$
$$3k = -25k + 20$$
$$28k = 20$$
$$k = \frac{20}{28} = \frac{4}{7}$$

Therefore, the correct answer is $\frac{4}{7}$, which you can also enter into the grid as .571.

Solving rational equations that become quadratics

In some cases, a rational equation may turn into a quadratic equation. Here is an example:

EXAMPLE

If p and q are both solutions to the equation $\frac{2}{x+4} = x+3$, then what is the value of pq?

To begin, multiply both sides of this equation by $x+4$ and simplify:

$$\frac{2}{x+4}(x+4) = (x+3)(x+4)$$
$$2 = x^2 + 7x + 12$$

To solve this quadratic equation, subtract 2 from both sides and then factor:

$$0 = x^2 + 7x + 10$$
$$0 = (x+2)(x+5)$$

Thus, $x = -2$ and $x = -5$. These are the two values of p and q, so $pq = 10$.

Solving Inequalities

An *inequality* is a mathematical statement that's similar to an equation, but which swaps out the equals sign (=) for one of the four inequality signs:

>> Less than: <

>> Greater than: >

>> Less than or equal to: ≤

>> Greater than or equal to: ≥

In this section, I solidify a few basic points about solving inequalities.

Using equation-solving skills with inequalities

Many students find inequalities somewhat confusing. But most of what you know about solving equations carries over when solving inequalities. Here is an example:

EXAMPLE

$$9x - 13 \leq 2x + 8$$

Which of the following is the solution set for this inequality?

(A) $x \leq 3$

(B) $x \geq 3$

(C) $x \leq -3$

(D) $x \geq -3$

To solve this inequality, use the same steps that you would when solving an equation:

$$9x - 13 \leq 2x + 8$$
$$7x - 13 \leq 8$$
$$7x \leq 21$$
$$x \leq 3$$

Therefore, Answer A is correct.

Knowing when to flip the inequality sign

Here's the key difference between solving equations and inequalities: When you multiply or divide an inequality by a negative number, you flip the inequality sign, as shown in Table 4-1.

TABLE 4-1 **Flipping Inequality Signs**

Before Multiplying or Dividing by a Negative Number	After Multiplying or Dividing by a Negative Number
<	>
>	<
≤	≥
≥	≤

EXAMPLE

Here is an example:

$$4x - 19 > 6x - 9$$

Which of the following is the solution set for this inequality?

(A) $x < 5$

(B) $x > 5$

(C) $x < -5$

(D) $x > -5$

To solve this inequality, begin by using the same steps that you would when solving an equation:

$$4x - 19 > 6x - 9$$
$$-2x - 19 > -9$$
$$-2x > 10$$

Now, to finish solving, you need to divide both sides of the inequality by –2. When you do this, flip the inequality sign from > to <:

$$x < -5$$

Therefore, Answer C is correct.

Lining up inequalities to solve systems

Some SAT questions present you with a pair of inequalities and ask you to draw a conclusion based on them. In some cases, you can solve this type of question by lining up the inequalities. To see how this works, look at the following example:

EXAMPLE

$$x - 2 \leq y + 5$$
$$-y \geq 1$$

If these inequalities are both true, which of the following is also true?

(A) $x \leq 6$

(B) $x \geq 6$

(C) $x \leq -6$

(D) $x \geq -6$

To answer this question, you want to create an inequality that has x on the left side, so change the first inequality by adding 2 to both sides as follows:

$$x - 2 \leq y + 5$$
$$x \leq y + 7$$

Now, your goal is to get the left side of the second inequality equal to $y + 7$, so that you can line up the two inequalities. To do this, begin by multiplying both sides by −1, flipping the inequality sign.

$$-y \geq 1$$
$$y \leq -1$$

Next, add 7 to both sides:

$$y + 7 \leq 6$$

Now, the inequality signs are both pointed in the same direction, so you can combine the two inequalities as follows:

$$x \leq y + 7 \leq 6$$

This inequality implies that $x \leq 6$ is also true, so Answer A is correct.

IN THIS CHAPTER

» **Understanding linear functions as words, tables, graphs, and equations**

» **Converting equations between slope-intercept form $y = mx + b$ and standard form $ax + by = c$**

» **Using slope, y-intercept, and points on the xy-graph to answer SAT questions about linear functions**

» **Solving parallel and perpendicular line problems**

» **Using linear functions to solve word problems**

Chapter **5**

Linear Functions

I f I could only discuss one topic to help you improve your SAT math score, it would have to be linear functions — that is, equations that have the form $y = mx + b$.

By my count, from 6 to 12 questions rely on your knowledge of linear functions. That's about 10% to 20% of the 58 questions on the test!

Fortunately, linear functions are an Algebra I topic, making them way easier to work with than quadratics, exponentials, or any other function covered in high school math. Unfortunately, if you're like most SAT students, you probably completed your Algebra I course as many as 2, 3, or even 4 years ago.

In this chapter, I provide a refresher on this all-important topic. To begin, I show you how to understand linear functions in four different ways: in words, tables, graphs, and equations. From there, you proceed to the main event, which is the slope-intercept form $y = mx + b$. You get clear on how to identify the y-intercept, the slope, and the input and output variables of a line.

I also show you how to work with a linear equation in standard form $ax + by = c$, and how to convert an equation to and from slope-intercept form. After that, you work with two formulas for finding the slope of a line on the xy-graph. Then, you learn a set of key strategies for discovering the equation of a line given a variety of parameters, such as slope, y-intercept, or one or more points on that line.

Next, you work with both parallel and perpendicular lines on the *xy*-graph. To finish up, I show you how to answer a variety of SAT word problems involving linear functions.

Interpreting Linear Functions as Words, Tables, Graphs, and Equations

Linear functions are very useful for understanding real-world events. That's one reason why the SAT places so much emphasis on them. In this section, I show you how linear functions arise in a variety of situations, and how to interpret them in four important ways:

>> Words

>> Tables

>> Graphs

>> Equations

After that, you'll be ready to answer a variety of SAT questions that depend on this understanding.

Understanding linear functions in four complementary ways

Linear functions help make sense of life situations that are very common, both on the SAT and elsewhere. For example, suppose your rich aunt, or somebody equally nice, wants to help you start a college fund. In this section, I show you four different ways to track the amount of money that she plans to give you.

Making words count

One way to make sense of numbers is by using *words* to describe them. For example, suppose your aunt tells you that she's opened a bank account for you with an initial balance of $2,000. Even better, she promises to deposit an additional $300 into the account every month.

This verbal description of your aunt's gift allows you to project how much money you'll have at any time in the future. For example, if you want to know how much money you'll have after three months, you simply count:

Open + 1st month + 2nd month + 3rd month
$2,000 + $300   + $300   + $300   = $2,900

Not bad for just being nice and never, ever forgetting to send her a birthday card!

Tabling your progress

A second way to understand your aunt's gift is by making a table (see Table 5-1 for an example). This is only slightly more formal than the count I made in the previous section.

TABLE 5-1 **Tracking the Money in Your Bank Account**

x = Months	y = Dollars in Account
0	2,000
1	2,300
2	2,600
3	2,900
4	3,200
5	3,500

Here, I've introduced the variables x and y to keep track of the numbers. The variable x stands for the number of months, and the variable y stands for the dollar amount in the account. Note that $x = 0$ tracks the *initial* value of $2,000 that your aunt used to open the account.

Graphing the results

A third way to clarify how your money will accrue is by making a *graph*. Figure 5-1 shows a graph of the information pulled from the table in the previous section.

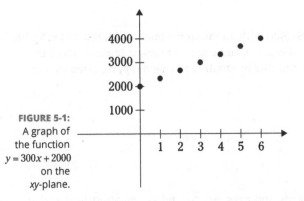

FIGURE 5-1:
A graph of
the function
$y = 300x + 2000$
on the
xy-plane.

A graph can give you a visual sense of how to interpret numbers. In this graph, for example, time proceeds on the x-axis from left to right. And the y-axis shows dollar values increasing in the upward direction. As you can see, the dots form a line that's increasing at a constant rate from one month to the next, indicating an increasing balance in your account.

Creating an equation (or function)

A fourth way to track numbers is by creating an *equation* to show the relationship between the month (x-value) and the dollar amount (y-value). When you can *input* an x-value and reliably *output* a single y-value, this type of equation is called a *function*.

The function to track the amount of money in your bank account on a month-by-month basis is:

$$y = 300x + 2000$$

This function of x includes two basic components that reliably produce a y-value:

» An *incremental value* $300x$, which represents the total amount of money placed in the account every month after it was opened.

» A *starting value* (or *initial value*) of 2000, which represents the amount of money that was used to open the account.

Remember that in a function, the incremental value is always attached to a variable such as x, and the starting value is always a constant — that is, a number by itself.

The general form of this function is $y = mx + b$, where m is the incremental value and b is the starting value. Later in this chapter, you answer SAT questions that are related to this type of function.

Answering SAT questions about linear functions

When you understand how linear functions can be expressed in four different ways — words, tables, graphs, or equations — you're ready to answer a variety of SAT questions. Many SAT questions present information to you in one of these four ways and ask you to interpret it in another way. For example:

EXAMPLE

Melanie agreed to buy her friend's car for $4,500 with an initial payment of $1,200 and monthly payments of $150 until the car is payed off. Assuming there are no taxes or fees, which of the following equations can be used to find the number of monthly payments, p, required to complete the purchase?

(A) $4500 = 150p + 1200$

(B) $4500 = 150p - 1200$

(C) $4500 = 1200p + 150$

(D) $4500 = 1200p - 150$

This question describes the function in words and asks you to find an equation to describe it. Here, the incremental value, 150, represents the monthly payment. This value is attached to the variable p, which rules out Answers C and D. And the starting value, $1,200, represents the initial payment. This value should be positive, because it helps to add up to the total amount of $4,500, which rules out Answer B. Therefore, Answer A is correct.

Here's another problem that presents you with information in one form and asks you to change it to a different form:

EXAMPLE

x	y
0	9
1	5
2	1
3	-3

This table shows a linear relation between four x-values and their corresponding y-values. Which of the following functions expresses the same relation?

(A) $y = 4x - 9$

(B) $y = -4x + 9$

(C) $y = 9x + 4$

(D) $y = -9x - 4$

This time, the problem presents information in a table and asks you to find the equation. Here, the starting value is 9, because this value is associated with an x-value of 0. And the incremental value is -4, because this is how much the value changes every time you add 1 to the x-value. Therefore, the equation is $y = -4x + 9$, so Answer B is correct.

As another example, here's a question that asks you to turn information from a table into information in words:

EXAMPLE

x = Hours	y = Total Amount Charged
1	$75.00
2	$105.00
3	$135.00
4	$165.00

When Dan works as a handyman, his total charge to the customer is based on an hourly rate plus a set amount. This table shows how much he charges for jobs that range from 1 hour to 4 hours. There is a linear relationship between the number of hours Dan works and the amount that he charges. Which of the following accurately reflects what the value $40 represents in the table?

(A) The amount that Dan charges per hour.

(B) The amount that Dan charges for the first hour that he works.

(C) The amount that Dan charges for each hour after the first hour that he works.

(D) The amount that Dan charges irrespective of the number of hours that he works.

According to the table, Dan charges $75 for the first hour. He charges $105 for two hours, which is $40 more than $75, so he charges $40 for the second hour. The problem states that there is a linear relationship among the numbers, so Dan charges $40 for each hour after the first hour that he works. Therefore, Answer C is correct.

Linear Function Basics

In the previous section, I show you how linear functions arise from a variety of real-world scenarios. In this section, you work with this function in a variety of ways that are more abstract. First, you use the slope-intercept form of the function, $y = mx + b$, to understand how changes to the values m and b affect the graph of the function. After that, you work with the standard form of the function, $ax + by = c$.

Slope-intercept form

The most common and useful form of a linear function is called *slope-intercept form*:

$$y = mx + b$$

In practice, the m and b are usually replaced by specific values and retain these values as *constants* in a given context — that is, the whole time you're working on a single problem. For example, the first column of Table 5-2 gives four different linear functions, each with different values of m and b.

TABLE 5-2 **Variety of Linear Functions and Three Coordinate Pairs for Each Function**

Function	Three coordinate pairs (x, y) that satisfy the function		
$y = 3x + 2$	(1, 5)	(2, 8)	(3, 11)
$y = 10x - 5$	(0, –5)	(10, 95)	(20, 195)
$y = 1,000x + 2,000$	(6, 8,000)	(12, 14,000)	(24, 26,000)
$y = -80x + 120$	(0, 120)	(1, 40)	(2, –40)

In contrast, within a single problem, the variables x and y are often assigned *coordinate pairs* (x,y) of values that make a specific equation true. For example, in Table 5-2, each remaining column provides a different coordinate pair that satisfies the function in that row.

In this section, I break the equation $y = mx + b$ into these four components to help make sense of them.

b is the y-intercept

In the equation $y = mx + b$, the constant b is the y-intercept of the graph — that is, the point where the graph intersects the y-axis. The coordinates of this point are $(0,b)$.

In word problems involving linear functions, the y-intercept b is always the starting value, or initial value, of the function. For example, suppose you begin a hiking trip at a 3,500-foot elevation and plan to hike up a mountain at a rate of 175 feet per hour. You can use the equation $y = 175x + 3500$ to calculate your elevation x hours after you begin hiking. As you can see, the number 3,500 is the elevation that you start with at time 0 — that is, at the start of the hike.

m is the slope

In the equation $y = mx + b$, the constant m is the slope of the graph — that is, the constant rate at which y either increases or decreases.

In word problems involving linear functions, the slope refers to the incremental value, or rate of change, depending on the value of x. For example, if a company has 50 employees and is planning to hire 10 new people per year, you can use the function $y = 10x + 50$ to calculate the projected number of employees in x years. Here, the value 10 is the rate of new hires per year.

x is the input value

In the equation $y = mx + b$, the variable x is the input value — that is, the number that is inputted into the equation in order to yield an output value of y.

In word problems, x is often a measurement of time. For example, suppose a tutor uses the equation $y = 35x + 15$ to calculate their hourly rate. In this case, x equals the number of hours they work. So, if the tutor works for 2 hours, you could calculate what they charge as follows:

$$y = 35(2) + 15 = 70 + 15 = 85$$

So they charge $85 for a 2-hour session.

y is the output value

In the equation $y = mx + b$, the variable y is the output value — that is, the value that's calculated for every possible x value inputted into the function.

In word problems, y is usually the final value that you're trying to calculate in a problem. For example, suppose that an office building under construction has currently reached a height of 17 floors, and construction is proceeding at a rate of 6 floors per month. You can use the equation $y = 6x + 17$ to calculate the number of floors that will be completed in x months. So, if you want to know how many floors will be completed in 4 months, plug in 4 for x into the function and calculate y:

$$y = 6(4) + 17 = 24 + 17 = 43$$

So the value y tells you that 43 floors will be completed in 4 months.

Standard form

The equation $ax + by = c$ is an alternative form of the linear equation and is called *standard form*. In most cases, it's less useful than slope-intercept form $y = mx + b$. However, because it shows up on the SAT with some frequency, you need to be able to work with it.

In this section, I show you how to change a standard-form linear equation into slope-intercept form, which can help you answer a variety of SAT questions. I also show you how to plot a line on the xy-graph directly from a standard-form equation.

Changing standard form to slope-intercept form

To answer many SAT questions, the best thing you can do when presented with a linear equation in standard form is to change it to slope-intercept form. To do this, follow these steps:

1. **Subtract over the x-term to the right side of the equation, placing it ahead of the constant term.**

2. **Divide every term in the equation by the coefficient of the y-term.**

For example:

What is slope of the linear equation $3x - 5y = 7$?

Begin by subtracting over the x-term $3x$:

$$-5y = -3x + 7$$

Notice in this step that I place the resulting x-term ahead of the constant 7. Now, divide every term of the equation by the coefficient of the y-term, which is -5. I do this in two steps:

$$\frac{-5y}{-5} = \frac{-3x}{-5} + \frac{7}{-5}$$
$$y = \frac{3}{5}x - \frac{7}{5}$$

The result is an equivalent linear equation in slope-intercept form, so the slope is $\frac{3}{5}$, which you can write equivalently as .6.

Converting slope-intercept form to standard form

To convert a linear function in slope-intercept form to standard form, follow these steps:

1. Subtract over the *x*-term to the left side of the equation, placing it ahead of the *y*-term.

2. If the resulting equation includes fractions, multiply each term by the least common denominator.

3. If the *x*-term is negative, multiply every term by –1 to flip the signs of every term in the equation.

For example:

EXAMPLE

Which of the following equations is the standard-form equivalent to the linear function $y = \frac{2}{3}x - \frac{3}{4}$?

(A) $8x + 12y = 9$

(B) $8x + 12y = -9$

(C) $8x - 12y = 9$

(D) $8x - 12y = -9$

To begin, subtract $\frac{2}{3}x$ from both sides of the equation:

$$-\frac{2}{3}x + y = -\frac{3}{4}$$

The resulting equation has fractions, so multiply each term by the lowest common denominator of these fractions, which is 12. I do this in two steps:

$$-\frac{2}{3}x(12) + y(12) = -\frac{3}{4}(12)$$
$$-8x + 12y = -9$$

This equation has a negative *x*-term, so multiply every term by –1, flipping the sign of every term. Again, I do this in two steps:

$$-1(-8x + 12y = -9)$$
$$8x - 12y = 9$$

Therefore, Answer C is correct.

Finding the intercepts of a standard-form linear equation

One advantage of a standard-form linear equation is that its *x*-intercept and *y*-intercept are easy to find.

1. To find the *x*-intercept, let $y = 0$ and solve for *x*.

2. To find the *y*-intercept, let $x = 0$ and solve for *y*.

An example should be helpful here.

Which of the following statements is true about the equation $-2x + 5y = -10$?

(A) The x-intercept and y-intercept are both positive.

(B) The x-intercept is positive and the y-intercept negative.

(C) The x-intercept is negative and the y-intercept is positive.

(D) The x-intercept and y-intercept are both negative.

To find the x-intercept, let $y = 0$ and solve for x:

$$-2x + 5(0) = -10$$
$$-2x = -10$$
$$x = 5$$

Thus, the x-intercept is 5, which is positive, ruling out Answers C and D.

To find the y-intercept, let $x = 0$ and solve for y:

$$-2(0) + 5y = -10$$
$$5y = -10$$
$$y = -2$$

Thus, the y-intercept is -2, which is negative, so Answer B is correct.

Graphing a linear equation in standard form

One way to graph a linear equation in standard form is to change it to slope-intercept form before graphing it. Another way is to find its x-intercept and y-intercept, as I show you in the previous section, and use these intercepts to graph the equation. For example:

Which of the following could be the graph of the equation $-3x + 4y = 12$?

(A)

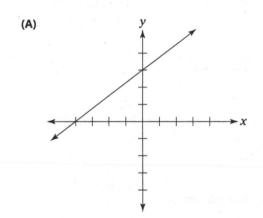

(B)

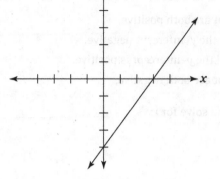

(C)

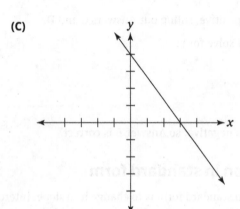

(D)

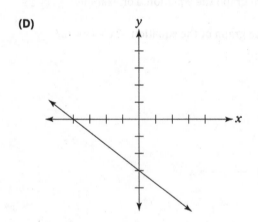

To begin, find the two intercepts of the equation:

$$-3x + 4(0) = 12 \qquad -3(0) + 4y = 12$$
$$-3x = 12 \qquad\qquad 4y = 12$$
$$x = -4 \qquad\qquad\quad y = 3$$

Now, plot these two intercepts on the xy-graph as $(-4,0)$ and $(0,3)$:

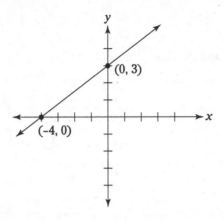

Therefore, Answer A is correct.

Using Two Formulas to Find the Slope

Many SAT questions are much easier to answer when you know how to find the slope of a linear function. There are two handy formulas for finding the slope, one useful for finding the slope of a graph and the other for finding the slope when you have two points on a line. In this section, I show you both of these important formulas.

Measuring the slope of a graph

When you have the graph of a line on the *xy*-plane, you can find the slope using the following formula:

$$\text{Slope} = \frac{\text{Rise}}{\text{Run}}$$

This formula uses two values that you can find by looking at the graph of a line:

>> The *rise* is how much the graph goes *up* or *down*.

>> The *run* is how much the graph goes *over toward the right*.

WARNING

Although the rise can go up or down, the run is *always* measured from left to right — that is, in the positive direction.

An example should help make these ideas clear:

EXAMPLE

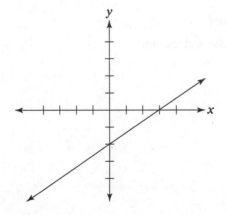

What is the slope of the line shown here?

To find the slope, measure both the rise and the run:

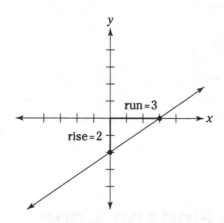

As you can see, here I've drawn a right triangle using two points on the graph, (0,−2) and (3,0). The rise is 2, and the run is 3, so input these values into the slope formula as follows:

$$\text{Slope} = \frac{\text{Rise}}{\text{Run}} = \frac{2}{3}$$

Therefore, the answer is $\frac{2}{3}$, which you can also write as .666 or .667.

Here's another example:

EXAMPLE

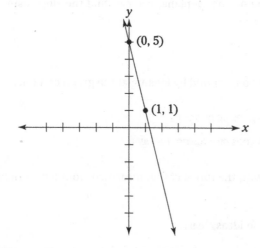

In this graph, what is the slope of the line?

Again, find the slope by measuring the rise and the run:

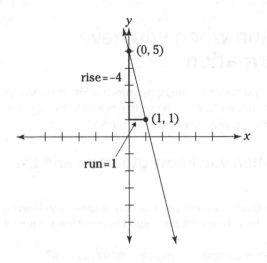

This time, I've drawn a right triangle using the points $(0,5)$ and $(1,1)$. In this case, the line slopes downward by 4, so the rise is -4, and the run is 1:

$$\text{Slope} = \frac{\text{Rise}}{\text{Run}} = \frac{-4}{1} = -4$$

Therefore, the answer is -4.

Measuring slope using two points

When you don't have the graph of a line, you can still find the slope of a line when you have the coordinates of two points on that line, (x_1, y_1) and (x_2, y_2). In this case, use the following formula:

$$m = \frac{y_2 - y_1}{x_2 - x_1}$$

What is the slope of a line that intersects the two points $(3, -7)$ and $(5, -2)$?

To answer this question, I let $(3, -7) = (x_1, y_1)$ and $(5, -2) = (x_2, y_2)$. (You can assign these points in either order, but make sure that you follow whatever order you choose.) Next, plug in the values for the four corresponding values into the formula:

$$m = \frac{-2 - (-7)}{5 - 3}$$

To find the slope, simplify this expression:

$$= \frac{-2 + 7}{5 - 3} = \frac{5}{2}$$

Therefore, the value is $\frac{5}{2}$, which you can also write as 2.5.

Solving Linear Function Problems

Many SAT questions about linear functions ask you to find the equation based upon information provided, such as slope, the y-intercept, or a point that's on the line. Others provide information about a related line that's either parallel or perpendicular to the line whose equation you're trying to find. In this section, I show you how to solve these types of questions.

Finding the equation when you have two pieces of information

You can answer a variety of SAT questions by plugging information from the problem into the slope-intercept form $y = mx + b$ and then finding the values m and b. In this section, I show you four common types of problems that appear frequently on the SAT.

Finding the equation when you know the slope and the y-intercept

The most straightforward type of question gives you both the slope and y-intercept. In this case, simply fill in the slope for m and the y-intercept for b into the equation $y = mx + b$. For example:

EXAMPLE

Which of the following functions has a slope of 2 and a y-intercept of -5?

(A) $y = 2x + 5$

(B) $y = 2x - 5$

(C) $y = -5x + 2$

(D) $y = -5x - 2$

In this problem, fill in 2 for m and -5 for b into the equation $y = mx + b$. Thus, the equation is $y = 2x - 5$, so Answer B is correct.

Answering point-slope questions

When an SAT question gives you the slope of a linear function and a point on that function, fill in the slope and the x and y coordinates into the equation $y = mx + b$ and solve for b. For example:

EXAMPLE

The graph of a line passes through the point $(4, -1)$ and has a slope of 3. Which of the following is the equation of that line?

(A) $y = 3x + 11$

(B) $y = 3x - 11$

(C) $y = 3x + 13$

(D) $y = 3x - 13$

Using the equation $y = mx + b$, plug in 3 for m, which gives you $y = 3x + b$. To find the value of b, also plug in 4 for x and -1 for y, and solve for b:

$$-1 = 3(4) + b$$
$$-1 = 12 + b$$
$$-13 = b$$

Therefore, the y-intercept of this equation is -13, so plug this value in for b into $y = 3x + b$:

$$y = 3x - 13$$

Therefore, Answer D is correct.

Finding the equation when you have the y-intercept and another point

When an SAT question gives you the y-intercept of a line and a point on that line, you can use a similar method: Fill in the y-intercept and the x and y coordinates into the equation $y = mx + b$ and solve for m. For example:

EXAMPLE

Which of the following could be the equation of a linear function that has a y-intercept of 5 and passes through the point $(6,-7)$?

(A) $y = 2x + 5$

(B) $y = -2x + 5$

(C) $y = \frac{1}{2}x + 5$

(D) $y = -\frac{1}{2}x + 5$

To solve the problem, plug in 5 for b into the equation $y = mx + b$. So the equation is $y = mx + 5$. Next, plug in 6 for x and -7 for y into this equation and solve for m:

$$-7 = 6m + 5$$
$$-12 = 6m$$
$$-2 = m$$

Thus, the slope is -2, so the equation is $y = -2x + 5$; therefore, Answer B is correct.

Solving two-point problems

You can find the slope of any linear function if you have two points on that function by using the slope formula I describe earlier in this chapter:

$$m = \frac{y_2 - y_1}{x_2 - x_1}$$

Here's an example:

EXAMPLE

If the equation of a line that intersects the points $(7,4)$ and $(-2,-14)$ is written in the form $y = mx + b$, what is the value of $|m + b|$?

To begin, plug in 4 and -14, respectively, for y_1 and y_2, and 7 and -2, respectively, for x_1 and x_2 into $m = \frac{y_2 - y_1}{x_2 - x_1}$; then solve for m:

$$m = \frac{-14 - 4}{-2 - 7} = \frac{-18}{-9} = 2$$

Thus, the slope of the equation is 2, so the equation is $y = 2x + b$. Now, using either point, plug the values of x and y into this equation and solve for b. I use the point $(7,4)$ here because the numbers are easier to work with:

$$4 = 2(7) + b$$
$$4 = 14 + b$$
$$-10 = b$$

Thus, the y-intercept of this line is -10, so the equation is $y = 2x - 10$. Therefore, $m = 2$ and $b = -10$, so:

$$|m + b| = |2 - 10| = |-8| = 8$$

Therefore, the correct answer is 8.

Understanding problems with parallel and perpendicular lines

When a pair of lines plotted on the xy-plane are parallel, their slopes are equivalent. And when a pair of lines are perpendicular, their slopes are negative reciprocals. You can use these two facts to answer more difficult SAT questions. In this section, I show you how.

Answering parallel line questions

A pair of parallel lines on the xy-plane have the same slope. You can use this information to answer SAT questions. For example:

The equation of line M is $y = \frac{1}{3}x - 2$. If line N is parallel to line M and includes the points $(-6, 5)$ and $(2, p)$, what is the value of p?

Lines M and N are parallel, so they have the same slope. Thus, the equation for line N is $y = \frac{1}{3}x + b$ for some b value. To find this value, plug in -6 for x and 5 for y into this equation and solve for b:

$$5 = \frac{1}{3}(-6) + b$$
$$5 = -2 + b$$
$$7 = b$$

Thus, the equation for line N is $y = \frac{1}{3}x + 7$. To find the value of p, plug in 2 for x and p for y into this equation and solve for p:

$$p = \frac{1}{3}(2) + 7 = \frac{2}{3} + 7 = 7\frac{2}{3} = \frac{23}{3}$$

Thus, the answer is $\frac{23}{3}$, which you can also write as 7.66 or 7.67.

Here's another problem that uses parallel lines on the xy-plane:

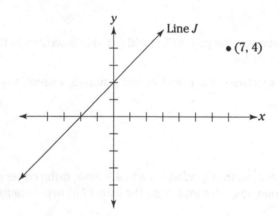

This figure shows the graph of line J. Line K (not shown) is parallel to line J and passes through the point $(7, 4)$. Which of the following is the equation of line K?

(A) $y = x - 1$

(B) $y = x - 3$

(C) $y = x - 5$

(D) $y = x - 7$

Measure the slope of line J as follows:

$$\text{Slope} = \frac{\text{Rise}}{\text{Run}} = \frac{1}{1} = 1$$

Line K is parallel to line J, so the slope of Line K is also 1; therefore, its equation is $y = x + b$. To find b, substitute 7 for x and 4 for y into this equation and solve for b:

$$4 = 7 + b$$
$$-3 = b$$

Thus, the y-intercept of line J is -3, so the equation is $y = x - 3$; therefore, Answer B is correct.

Using perpendicular lines to solve problems

REMEMBER

When a pair of lines on the xy plane are perpendicular — that is, when they intersect at a right angle — their slopes are negative reciprocals of each other. Table 5-3 shows a variety of values for the slope of a line and the slope of a line perpendicular to it.

TABLE 5-3 **Pairs of Negative Reciprocals**

Slope	Slope of Perpendicular
2	$-\dfrac{1}{2}$
$\dfrac{3}{4}$	$-\dfrac{4}{3}$
$-\dfrac{1}{5}$	5
$-\dfrac{2}{3}$	$\dfrac{3}{2}$

You can use this information to answer SAT questions that give you information lines on the xy-plane that are perpendicular to each other. For example:

EXAMPLE

When graphed on the xy-plane, what is the slope of a line whose graph is perpendicular to $y = -\dfrac{2}{7}x + 3$?

The slope of $y = -\dfrac{2}{7}x + 3$ is $-\dfrac{2}{7}$, so the slope of a perpendicular line is $\dfrac{7}{2}$, which you can also write as 3.5.

Here's a more difficult question that hinges on perpendicular lines:

EXAMPLE

If $f(x) = 4x + 5$ intersects at a right angle with $g(x)$ when $x = -1$, then $g(x) =$

(A) $-4x + \dfrac{3}{4}$

(B) $\dfrac{1}{4}x + \dfrac{5}{4}$

(C) $-\dfrac{1}{4}x + \dfrac{3}{4}$

(D) $-\dfrac{1}{4}x + \dfrac{5}{4}$

The graph of $f(x)$ has a slope of 4, so the graph of $g(x)$ has a slope of $-\dfrac{1}{4}$; thus, $g(x) = -\dfrac{1}{4}x + b$ for some value of b. The two lines intersect when $x = -1$, so plug this value into $f(x) = 4x + 5$:

$$f(-1) = 4(-1) + 5 = -4 + 5 = 1$$

Thus, the two lines intersect at the point $(-1, 1)$. Substitute -1 for x and 1 for $g(x)$ into $g(x) = -\dfrac{1}{4}x + b$ and solve for b:

$$1 = -\dfrac{1}{4}(-1) + b$$

$$1 = \dfrac{1}{4} + b$$

$$\dfrac{3}{4} = b$$

Thus, $g(x) = -\dfrac{1}{4}x + \dfrac{3}{4}$, and so Answer C is correct.

Solving Word Problems

Most students find word problems to be the most confusing on the SAT. In this section, you apply all the tools in this chapter to a variety of word problems that require knowledge of linear functions.

Angela's flowers

EXAMPLE

Angela makes colorful paper flowers to sell on her online store. She started the day with a stock of 32 flowers, and made new ones at a rate of 6 per hour, working for 8 hours. If $f(h)$ is a function that calculates the number of flowers that Angela had after working for h hours, which of the following correctly shows $f(h)$ and its proper application to calculate Angela's output of flowers for the day?

(A) $f(h) = 8h + 32$
$f(6) = 8(6) + 32$

(B) $f(h) = 6h + 32$
$f(8) = 6(8) + 32$

(C) $f(h) = 32h + 8$
$f(6) = 32(6) + 8$

(D) $f(h) = 32h + 6$
$f(8) = 32(8) + 6$

The correct function uses h as an input value representing hours, and $f(h)$ as an output value representing flowers:

$$f(h) = mh + b$$

Angela started the day with 32 flowers, so this value equals b:

$$f(h) = mh + 32$$

Thus, Answers C and D are ruled out. She worked at a rate of 6 flowers per hour for a total of 8 hours. Thus, you can substitute 6 for m in the function, and then substitute 8 for h when applying the function:

$$f(h) = 6h + 32$$
$$f(8) = 6(8) + 32$$

Therefore, Answer B is correct.

Peter's autos

EXAMPLE

Peter is an auto mechanic who charges a flat dollar amount plus an hourly rate for his services. If Peter charges $195 for 3 hours of service and $285 for 5 hours of service, what flat dollar amount does Peter charge for his services? (Disregard the dollar sign when gridding in your answer.)

Peter charges a flat dollar amount plus an hourly rate, so you can use the equation $y = mx + b$ to calculate this value, where m is his hourly rate and b is the flat dollar amount. He charges $195 for 3 hours of service, and $285 for 5 hours of service, so you can use the slope formula:

$$m = \frac{y_2 - y_1}{x_2 - x_1} = \frac{285 - 195}{5 - 3} = \frac{90}{2} = 45$$

Thus, the equation $y = 45x + b$ is true. Plug in 3 for x and 195 for y, and solve for b:

$$195 = 45(3) + b$$
$$195 = 135 + b$$
$$60 = b$$

Therefore, Peter charges a flat dollar rate of $60, so the answer is 60.

Jacqueline's ascent

EXAMPLE

Jacqueline lives on the 10th floor of her apartment building. For exercise, she often walks up to the 35th floor to visit a friend who lives there, walking at a constant rate of 20 seconds per floor. If Jacqueline uses the linear equation $f(t) = \frac{1}{20}t + 10$, using an input of $t = 0$ as her starting time at the base of the stairs on the 10th floor, and $f(t)$ as the floor that she is on at time t, how many seconds does Jacqueline take to reach the floor where her friend lives?

This question is tricky because there is a lot of information to assimilate. Begin by drawing a graph of $f(t) = \frac{1}{20}t + 10$ to get a visual sense of the problem:

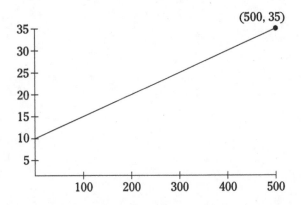

This graph charts Jacqueline's ascent from the 10th floor to the 35th floor. She takes 20 seconds to climb from one floor to the next, so she takes 100 seconds to climb 5 floors, and 500 seconds to climb the 25 floors from the 10th floor to the 35th floor, so the answer is 500.

IN THIS CHAPTER

» Solving systems of linear equations using substitution and elimination

» Discovering an easy trick to solve some systems of equations on the SAT

» Handling word problems in two variables

» Understanding systems with no solution or infinitely many solutions

» Working with systems of linear inequalities

» Solving systems of non-linear equations

Chapter **6**

Systems of Equations

One of the most common SAT math topics is systems of equations. In fact, of the 58 questions that you'll face on your SAT, 3 to 6 of them will most likely require you to work with a system of equations in one way or another.

In this chapter, you begin by exploring two ways of solving systems of linear equations: substitution and elimination. Then, I show you a trick for solving certain SAT questions quickly and easily.

After that, you encounter a classic type of word problem that is easiest to solve using a system of linear equations. I then explain why some systems of equations have no solution, while others have an infinite number of solutions, and I show you how to answer some of the most difficult SAT questions without too much trouble.

To finish up, you discover how to answer SAT questions that include systems of non-linear equations.

Systems of Linear Equations

A *system of linear equations* (also called *simultaneous equations*) is a set of equations that include more than one variable. On the SAT, you will most likely encounter only systems of two equations in two variables (usually x and y).

In this section, I show you how to solve systems of linear equations.

Solving systems of linear equations

On the SAT, a system of linear equations is most often presented as a pair of linear equations in standard form. Here is an example:

$$4x + y = 7$$
$$-3x + 5y = 12$$

In this section, I discuss two ways to solve systems of linear equations: substitution and elimination.

Substitution

A quick way to solve a system of linear equations is called *substitution*.

REMEMBER

To solve a system of linear equations using substitution, follow these steps:

1. Solve either equation for one variable in terms of the other.

2. Substitute (plug in) the resulting value for that variable into the other equation.

3. Solve that equation for the variable that remains.

4. Plug the result into either equation and solve for the remaining variable.

TIP

Substitution works best when at least one variable in the system you're working with has a coefficient of 1. Here is an example:

$$4x + y = 7$$
$$-3x + 5y = 12$$

EXAMPLE

In this system of equations, what is the value of xy?

1. Solve either equation for one variable in terms of the other.

In this case, the y-variable in the first equation has a coefficient of 1. To begin, solve this equation for y in terms of x:

$$y = -4x + 7$$

2. Substitute (plug in) the resulting value for that variable into the other equation.

Here, plug in $y = -4x + 7$ for y into the second equation:

$$-3x + 5(-4x + 7) = 12$$

3. Solve that equation for the variable that remains.

Begin by distributing and combining like terms, then solve for x.

$$-3x + -20x + 35 = 12$$
$$-23x + 35 = 12$$
$$-23x = -23$$
$$x = 1$$

4. Plug the result into either equation and solve for the remaining variable.

In this case, plugging in 1 for x into $4x + y = 7$ is easier:

$$4(1) + y = 7$$
$$4 + y = 7$$
$$y = 3$$

Thus, the solution to the system is $x = 1$ and $y = 3$, so $xy = 3$.

Elimination

Consider the following question:

$$5x - 2y = 3$$
$$-7x + 4y = -15$$

EXAMPLE

In this system of equations, what is the value of $\frac{y}{x}$?

This system doesn't include a variable with a coefficient of 1, so if you try to solve either equation for a variable, the result will be a rational expression. For example, look what happens when you try to solve the first equation for x:

$$5x - 2y = 3$$
$$5x = 2y + 3$$
$$x = \frac{2y + 3}{5}$$

If you plug this value into the second equation, the result will be so complicated that you might well make a calculation error.

When a system of linear equations has no variable with a coefficient of 1, a better way to solve it is with *elimination*.

To solve a system of equations using elimination, follow these steps:

REMEMBER

1. Multiply one or both equations by constants so that the resulting equations include two variable terms that will cancel out each other out when added together.

2. Add the two equations together to create a third equation in only one variable.

3. Solve that equation for the variable that remains.

4. Plug the result into either equation and solve for the remaining variable.

Take another look at the system of equations in the example question:

$$5x - 2y = 3$$
$$-7x + 4y = -15$$

If you multiply every term of the first equation by 2, here is the result:

$$10x - 4y = 6$$
$$-7x + 4y = -15$$

This system of equations is equivalent to the original system, but now the y-terms include the coefficients 4 and -4. Thus, if you add these two equations together, these two terms cancel each other out, as follows:

$$3x = -9$$

You can solve this equation easily:

$$x = -3$$

To complete the problem, plug in -3 for x into the easier of the two original equations:

$$5(-3) - 2y = 3$$
$$-15 - 2y = 3$$
$$-2y = 18$$
$$y = -9$$

So, the solution to the system of equations is $x = -3$ and $y = -9$, and so $\frac{y}{x} = \frac{-9}{-3} = 3$.

EXAMPLE

When using this method, in some cases you may need to multiply both equations by different values so that the two resulting equations include a pair of variables that will cancel each other out when added. Here is an example:

$$5x - 3y = -1$$
$$-3x + 2y = 1$$

In this system of equations, what is the value of $x^2 + y^2$?

In this case, you can't multiply either equation by any integer and expect one of the variable values to drop out when you add the equations together. Instead, multiply the first equation by 2 and the second equation by 3, as follows:

$$10x - 6y = -2$$
$$-9x + 6y = 3$$

Now, when you add the two values together, the result is an equation that includes x but not y:

$$x = 1$$

Plug this value back into either equation — I always try to choose the equation with numbers that are easier to work with:

$$-3(1) + 2y = 1$$
$$-3 + 2y = 1$$
$$2y = 4$$
$$y = 2$$

Therefore, the solution to the system is $x = 1$ and $y = 2$, so $x^2 + y^2 = 1 + 4 = 5$.

Looking for a fast solution

In some cases, an SAT question may offer an opportunity to solve a system of linear equations quickly and easily. For example, consider the following question:

EXAMPLE

$$3x - 4y = 410$$
$$4x - 3y = 570$$

What is the value of $7x - 7y$ in this system of equations?

You could solve this problem using the elimination method outlined earlier in this chapter. But a faster way appears when you notice that the expression $7x - 7y$ is the result when you add the two left-hand expressions in the system. Thus, you can answer the question by simply adding the two equations:

$$7x - 7y = 980$$

So the answer is 980.

Here's another example of a question that offers you a quick way to solve a system of equations. See if you can spot it in this example:

EXAMPLE

$$8x + 9y = 22$$
$$6x + 6y = 15$$

Given this system of equations, what is the value of $2x + 3y$?

Again, instead of diving in to the problem immediately, step back from it and notice that when you subtract the top equation minus the bottom, the result is as follows:

$$2x + 3y = 7$$

So the answer is 7.

Here's a final example that's a little more difficult than the previous two problems. See if you can find a quick way to do it:

EXAMPLE

$$10x + 20y = 22$$
$$5x - 5y = 9$$

In this system of linear equations, what does $25x + 35y$ equal?

The trick here is to see if you can build the expression $25x + 35y$ using some combination of the expression on the left side of the equation. A little experimentation will lead you to notice that if you multiply the first equation by 2, the result is as follows:

$$20x + 40y = 44$$
$$5x - 5y = 9$$

Now, when you add the two equations, the result gives you the answer:

$$25x + 35y = 53$$

So the answer is 53.

On the SAT, a system of three equations can almost certainly be solved using this type of trick. Here is an example:

EXAMPLE

$$x + y + z = 10$$
$$2x + 3y + 4z = 30$$
$$7x + 8y + 9z = 50$$

In this system of equations, what is the value of $5x + 6y + 7z$?

WARNING

If you tried to solve this system of equations using normal methods, such as substitution and elimination, the question would take so long to answer that you'd probably miss half the other questions on the test!

So, if you're going to work on this problem at all, look for a clever way to get the answer. To begin, try adding the three equations:

$$10x + 12y + 14z = 90$$

Now, compare the left side of this equation to the expression you're trying to evaluate. Notice that $5x + 6y + 7z$ is exactly half of $10x + 12y + 14z$. So, if you divide both sides of this equation by 2, the result is as follows:

$$5x + 6y + 7z = 45$$

Therefore, the answer is 45.

You won't be able to solve every system of equations you see on the SAT in this manner, but keep an eye out for opportunities like these!

Solving word problems with a system of linear equations

In this section, I show you two ways to use a system of linear equations to solve word problems. The first method uses a chart to organize information, and the second method allows you to solve the problem without using a chart.

For students who love charts

Some students — especially students who are very visually oriented — find that a chart can be helpful when solving certain types of word problems. For example, consider this classic word problem:

EXAMPLE

Genice earned $144.50 for her soccer team by selling brownies and chocolate chip cookies after school. She sold each brownie for $2.00 and each cookie for $1.50. If she sold a total of 81 items, how many cookies did she sell?

Begin by declaring two variables:

b = the number of brownies Genice sold

c = the number of cookies Genice sold

In Table 6-1, I've organized this information into a chart.

TABLE 6-1 ## Table for Solving a System of Equations Word Problem

Type of item	Number of items	Cost per item	Total earnings
Brownies	b	$2.00	$2b$
Cookies	c	$1.50	$1.5c$
Total	81	N/A	144.5

Genice sold b brownies at $2.00 each, so her total earnings for brownies was $2.00b, which simplifies to $2b$. Similarly, she sold c cookies at $1.50 each, so she earned a total of $1.50c, which simplifies to $1.5c$.

Additionally, I've placed the totals in the third row of the chart: She sold a total of 81 items and earned a total of $144.50, which simplifies to 144.5.

To set up the two equations for the problem, simply add the two columns that have a number in the third row:

$$b + c = 81$$
$$2b + 1.5c = 144.5$$

To solve this system by elimination, multiply the first equation by -2:

$$-2b - 2c = -162$$
$$2b + 1.5c = 144.5$$

Now, when you add these two equations, the variable b drops out:

$$-0.5c = -17.5$$

To solve, multiply both sides by -2:

$$c = 35$$

Therefore, Genice sold 35 cookies.

For students who hate charts

If you don't like charts or find them unhelpful, you can still solve word problems that require you to set up a system of equations. Here is an example:

EXAMPLE

Antoine sold tickets for the opening night performance of his school play. Tickets cost $5.00 for adults and $3.50 for children under 12 years old. Altogether, Antoine sold 106 tickets for a total of $467.00. How many children's tickets did he sell?

As in the last example, begin by declaring two variables:

a = the number of adults' tickets that Antoine sold

c = the number of children's tickets that Antoine sold

Thus, $a + c$ represents the total number of tickets sold:

$$a + c = 106$$

Additionally, $5.00a$ represents the number of dollars that Antoine made on adults' tickets, and $3.50c$ represents the number of dollars that he made on children's tickets. So, adding these two values gives you his total ticket revenue:

$$5a + 3.5c = 467$$

Together, these two equations compose the system that you need to solve. Begin by multiplying the first equation by -5:

$$-5a + -5c = -530$$
$$5a + 3.5c = 467$$

When you add these two equations, the variable a drops out:

$$-1.5c = -63$$

To solve, divide both sides of the equation by -1.5:

$$c = 42$$

Therefore, Antoine sold 42 children's tickets.

In some cases, especially on the calculator section of the SAT, a question may ask you to set up a system of equations for a word problem without actually solving it. Questions like these may seem long and daunting, but because you don't have to actually solve them, they may not be as difficult as you think. Here is an example:

EXAMPLE

A plumbing contractor orders both ¾-inch copper pipe and 1½-inch PVC pipe for a job. The copper pipe costs $3.65 per foot, while the PVC pipe costs $1.80 per foot. The contractor orders a total of 760 feet of pipe and pays a total of $2,367.00 for it. If C represents the number of feet of copper pipe purchased, and P represents the number of feet of PVC pipe purchased, which of the following systems of equations could be used to find the amounts that the contractor spent on each type of pipe?

(A)
$$1.8C + 3.65P = 730$$
$$C + P = 2,367$$

(B)
$$1.8C + 3.65P = 2,367$$
$$C + P = 760$$

(C)
$$3.65C + 1.8P = 760$$
$$C + P = 2,367$$

(D)
$$3.65C + 1.8P = 2,367$$
$$C + P = 760$$

This problem looks discouraging because it's long and tedious, with terminology about plumbing supplies you've probably never heard of before. But at its core, it's actually easier than the questions about baked goods and tickets because you don't have to solve the equations, just set them up.

Begin by noticing that C and P stand, respectively, for the number of feet of copper and PVC pipe that the contractor ordered. This total number of feet $(C + P)$ adds up to 760, so you can rule out Answers A and C. Next, note that C (copper pipe) costs $3.65 and that P (PVC pipe) costs $1.80, so these values in the first equation need to be paired correctly. This rules out Answer B, so Answer D is correct.

Working with Problematic Systems of Equations

So far, each of the systems of linear equations you've worked with in this chapter have had exactly one solution. Some SAT questions, however, ask you to grapple with systems that have no solution or an infinite number of solutions.

In this section, you examine these three types of systems to see what makes them all different from each other. Then, you discover how to answer some of the most conceptually difficult questions on the SAT without breaking a sweat!

Understanding systems with one solution

You can always graph a system of linear equations that has one solution as a pair of intersecting lines. For example, Figure 6-1 contains the graph of the following system.

$$2x + y = 4$$
$$-x + y = 1$$

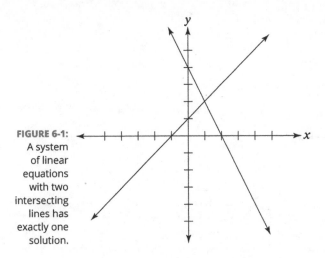

FIGURE 6-1:
A system
of linear
equations
with two
intersecting
lines has
exactly one
solution.

In this case, the two lines intersect at the point (1,2), the solution to this system of equations. Every system of linear equations that has a single solution can be graphed similarly, as a pair of intersecting lines on the xy-plane.

Troubleshooting problematic systems

Problematic systems of linear equations come in two varieties:

» Systems with no solutions

» Systems with infinitely many solutions

Each of these problems stems from the same source. When you understand how these problems arise, you'll be ready to answer a difficult variety of SAT question. In this section, I show you how and why problems occur in systems of linear equations.

Systems with no solutions

Consider the following system of linear equations:

$$2x + y = 4$$
$$2x + y = 5$$

This system of equations has no solution. This fact makes sense when you examine the two equations and realize that no possible values of x and y could make $2x + y$ equal both 4 and 5.

Figure 6-2 helps to illuminate where the problem lies. When graphed on the xy-plane, these two equations are a pair of parallel lines, which never intersect, so the system has no solution.

You can camouflage this problem by multiplying each equation by a different value. Here is an example:

$$4x + 2y = 8$$
$$6x + 3y = 15$$

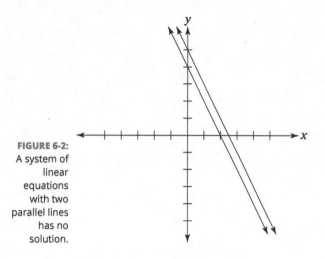

FIGURE 6-2:
A system of
linear
equations
with two
parallel lines
has no
solution.

Here, I've multiplied the first equation by 2 and the second equation by 3. This is essentially the same system of equations with the same problem in that it has no solution. But this version of the system highlights that the core problem with this system of equations is as follows:

>> The coefficients of the two *x*-values are *proportional* to the coefficients of the *y*-values (in this case, they are in a 2:3 ratio).

>> The two constants (8 and 15) are *not proportional* to those coefficients.

As a result of these facts, the system has no solution.

Systems with infinite solutions

Now, consider the following system of linear equations:

$$2x + y = 4$$
$$2x + y = 4$$

In a sense, this isn't really a system of equations, but rather a single equation with an infinite number of solutions. As you can see, every possible *x*-value has a corresponding *y*-value that could make the expression $2x + y$ equal 4.

Figure 6-3 makes visual sense of this situation. Because this system has only one distinct equation, the result when graphed on the *xy*-plane is simply a single line. You can think of this as a line on top of the same line, both of which "intersect" at *every* point.

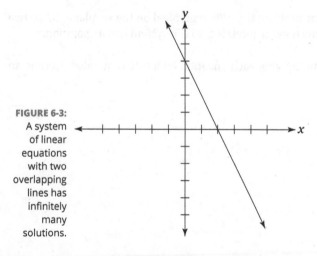

FIGURE 6-3:
A system
of linear
equations
with two
overlapping
lines has
infinitely
many
solutions.

As in the previous example, you can make this system look normal by multiplying each equation by a different value:

$$10x + 5y = 80$$
$$-8x - 4y = -60$$

Here, I've multiplied the first equation by 5 and the second equation by −4. The system looks normal, but it has the same problem as before. This version of the system shows that the core problem is as follows:

>> The coefficients of the two x-values are *proportional* to the coefficients of the y-values (in this case, they are in a 5:−4 ratio).

>> The two constants (80 and −60) are *also proportional* to those coefficients.

As a result of these facts, the system has infinitely many solutions.

Understanding what makes a system problematic

The key factor that causes problems to arise with a system of equations is *proportionality* between the x and y coefficients.

When the x and y coefficients of the two equations are proportional to each other, the system can never have a single solution. If the constants are not in the same proportion, the system has no solution; alternatively, if the constants are in the same proportion, the system has an infinite number of solutions.

Table 6-2 summarizes this situation.

TABLE 6-2 **Systems of Equations with 0, 1, or Infinite Solutions**

Number of Solutions	Coefficients of x and y	Constants
One solution	Not proportional	N/A
No solution	Proportional	Not proportional with x and y
Infinite solutions	Proportional	Proportional with x and y

Answering SAT questions about problematic systems

With the insights of the previous section in mind, consider the following SAT question:

$$4x + 7y = 3$$
$$12x + ny = 11$$

EXAMPLE

In this system of equations, what value of n results in a system with no solution?

When a system of linear equations has no solution, the coefficients of the x and y variables in the two equations must be in proportion to each other. Here, the ratio of the x variables in the first and second equations is 1:3. Thus, the ratio of the coefficients 7 and n must be 1:3. Thus, $n = 21$.

As another example, consider this question:

EXAMPLE

$$2x - 5y = k$$
$$-8x + 20y = -24$$

What value of k in this system of equations produces infinitely many solutions?

In this case, the x and y coefficients in the first and second equations are in a $1:-4$ ratio, respectively. Thus, if k and -24 are also in a $1:-4$ ratio, the system will have infinitely many solutions. Therefore, $k = 6$.

Here's one final question that would stump most SAT test-takers:

EXAMPLE

$$1.7x - 0.6y = b$$
$$5.1x - ay = 1.2$$

If this linear system of equations results in infinitely many solutions, what is the value of $a + b$?

In this system, the ratio of the x-coefficients in the first and second equations is $1:3$. Thus, the ratio of 0.6 and a is also $1:3$, so $a = 1.8$. Similarly, the ratio of b to 1.2 is $1:3$, so $b = 0.4$. Therefore, $a + b = 1.8 + 0.4 = 2.2$, so this is the correct answer.

Systems of Linear Inequalities

A system of linear inequalities is similar to a system of linear equations, but the equals signs are replaced by inequality symbols ($<$, $>$, $\leq$, or $\geq$). Here is an example:

$$x + y \geq 25$$
$$100x + 175y \geq 3500$$

In this section, I show you how to set up linear inequalities to solve word problems. I also show you how to graph linear inequalities.

Setting up systems of linear inequalities to solve word problems

Earlier in this chapter, I discuss how to set up and solve word problems using systems of linear equations. These skills also come into play when setting up a system of linear inequalities to solve a word problem.

However, when setting up a system of linear inequalities, you need to consider whether an expression can be *less than* or *greater than* a given value. Here is an example to help clarify this point:

EXAMPLE

Michele is an employment recruiter who earns 100 points when she fills a basic job slot and 175 points when she fills an elite job slot. Each month, she needs to fill at least 25 job slots and earn 3,500 or more points in order to meet her quota. If B and E represent, respectively, the number of basic and elite job slots that Michele fills in a given month, which of the following systems of inequalities could be used to represent the set of conditions in which she success-fully meets her quota?

(A)
$$B + E \leq 25$$
$$175B + 100E \leq 3500$$

(B)
$$B + E \leq 25$$
$$175B + 100E \geq 3500$$

(C)
$$B + E \geq 25$$
$$100B + 175E \leq 3500$$

(D)
$$B + E \geq 25$$
$$100B + 175E \geq 3500$$

Thinking about this problem as a system of linear equations yields the following result:

$$B + E = 25$$
$$100B + 175E = 3500$$

This rules out Answers A and B. The number of job slots that Michele needs to fill is *greater than* 25, and the number of points she earns is *greater than* 3,500, so you can rewrite these two equations as the following inequalities:

$$B + E \geq 25$$
$$100B + 175E \geq 3500$$

Thus, Answer D is correct.

Graphs with linear inequalities

Another common type of SAT question asks you to identify a system of linear inequalities as it appears on the *xy*-plane. This type of question usually asks you to decide which shaded region is correct.

When a linear inequality is expressed in slope–intercept form, the direction of the inequality sign determines whether the region above or below the line should be shaded:

>> When $y \leq mx + b$, the region *below* the line should be shaded.

>> When $y \geq mx + b$, the region *above* the line should be shaded.

Here is an example:

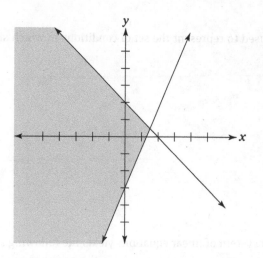

This graph represents which of the following systems of linear equalities?

(A) $y \le 2x - 3$
$y \le -x + 2$

(B) $y \le 2x - 3$
$y \ge -x + 2$

(C) $y \ge 2x - 3$
$y \le -x + 2$

(D) $y \ge 2x - 3$
$y \ge -x + 2$

The only differences among these answers are the inequality signs ($\le$ or $\ge$), so all four answers yield graphs that include identical lines on the xy-plane. The graph of the rising line is $y = 2x - 3$, because the slope is positive. Similarly, the graph of the falling line is $y = -x + 2$.

The region *above* the rising line is shaded, so this inequality is $y \ge 2x - 3$. And the region *below* the falling line is shaded, so this inequality is $y \le -x + 2$. Therefore, Answer C is correct.

When this type of question presents a system of linear inequalities in standard form ($ax + by \le c$ or $ax + by \ge c$), convert these inequalities to slope–intercept form. Here is an example:

REMEMBER

$-2x + y \le -1$
$3x - 4y \ge -8$

EXAMPLE

Which of the following graphs correctly represents this system of inequalities?

(A)

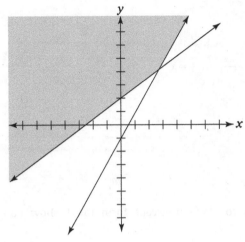

(B)

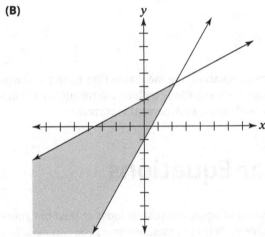

(C)

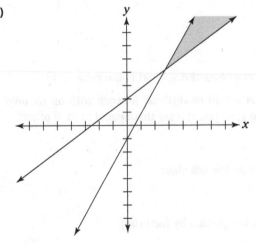

(D)

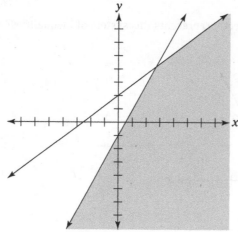

To begin, convert both equations to slope-intercept form (as I show you how to do in Chapter 4):

$$-2x + y \le -1 \qquad 3x - 4y \ge -8$$
$$y \le 2x - 1 \qquad -4y \ge -3x - 8$$
$$y \le \frac{3}{4}x + 2$$

Notice that when converting the second equation, the inequality flips from ≥ to ≥ when you divide both sides of the equation by −4. Now, both equations include the inequality ≤. Thus, the shaded region of the graph should be below both lines, so Answer D is correct.

Systems of Non-Linear Equations

When an SAT question involves a system of equations that includes at least one non-linear equation, you can't use elimination to solve it. In these cases, however, you can usually solve by substitution. This method often involves simply setting the two equations equal to each other. Here is an example:

$$y = 7x^2 + 3x + 18$$
$$y = 8x^2 + x - 6$$

EXAMPLE

Let $k > 0$ be a solution to this system of equations. What is the value of k?

Either of these quadratic functions would be difficult to work with on its own. However, both right-side expressions are equal to y, so you can set them equal to each other:

$$7x^2 + 3x + 18 = 8x^2 + x - 6$$

Next, subtract over all of the terms on the left side:

$$0 = x^2 - 2x - 24$$

You can solve the resulting quadratic equation by factoring:

$$0 = (x + 4)(x - 6)$$

Thus, the two solutions are $x = -4$ and $x = 6$. However, $k > 0$, so the correct answer is 6.

In some cases, you may need to do a little algebra to make this strategy workable. Here is an example:

EXAMPLE

$$y = x^2 - 6x + 5$$
$$4x + y = 8$$

When graphed on the xy-plane, these two equations intersect at (a,b) in Quadrant 2 of the coordinate plane. What does b equal?

In this system of equations, the variable y is already isolated in the first equation, so you can plug in $x^2 - 6x + 5$ for y in the second equation:

$$4x + x^2 - 6x + 5 = 8$$

Now, combine like terms and subtract 8 from both sides of the equation:

$$x^2 - 2x + 5 = 8$$
$$x^2 - 2x - 3 = 0$$

The result is a quadratic equation that you can solve by factoring:

$$(x + 1)(x - 3) = 0$$

Thus, $x = -1$ and $x = 3$. The solution (a,b) occurs in Quadrant 2 of the xy-plane, where the x-value is negative, so $a = -1$. You can find b by plugging in -1 for x and b for y into either equation, so use the second equation:

$$4x + y = 8$$
$$4(-1) + b = 8$$
$$-4 + b = 8$$
$$b = 12$$

Thus, the correct answer is 12.

3

Tackling Problem Solving and Data Analysis

IN THIS PART . . .

Solving problems using ratios, proportions, and percentages

Answering SAT questions that involve statistics and probability

Interpreting data presented in a variety of visual formats, such as tables, graphs, histograms, and scatterplots

IN THIS CHAPTER

» **Understanding ratios**

» **Using a ratio to set up a proportional equation**

» **Solving word problems using proportional equations**

» **Answering basic SAT questions involving percentages**

» **Solving tricky percent increase and percent decrease problems**

Chapter **7**

Ratios, Proportions, and Percentages

I n this chapter, you work with a variety of common ways to quantify relationships between numbers based on multiplication and division.

To begin, you discover how a ratio relates a pair of values in a way that's similar to a fraction. You then use ratios to set up proportional equations to answer SAT questions. With these tools in place, I show you how to solve a bunch of common SAT questions involving unit conversion, population density, similar triangles, and values that are directly proportional.

After that, the focus is on percentages. You answer a variety of SAT questions involving percentages by rewriting statements as equations and then solving them. Then, you review a common strategy for solving percent increase and decrease problems that you probably learned in school. I extend this knowledge by giving you a formula that you may find more helpful when answering SAT percent increase and decrease questions.

Understanding Ratios and Proportions

In this section, you use ratios — that is, fractional relationships between numbers — to set up proportional equations that you can use to answer SAT questions.

Understanding ratios

A *ratio* is a relationship between a pair of numbers based on division. You can use your knowledge of fractions to help you understand and work with ratios. For example:

EXAMPLE

Ms. Fisher's fifth grade class includes 12 children who walk to and from school and 15 children who take the school bus. What is the ratio of walkers to bus riders?

(A) 2:3

(B) 3:4

(C) 4:5

(D) 5:6

To solve the problem, make a fraction using the first quantity, 12, as the numerator and the second quantity, 15, as the denominator, then simplify this fraction (if needed, flip to Chapter 2 for a review of fractions):

$$\frac{12}{15} = \frac{4}{5}$$

The resulting fraction is analogous to the ratio 4:5, so Answer C is correct.

Creating proportional equations

A *proportional equation* includes a pair of fractions built from ratios, connected with an equals sign. You can use proportional equations to find and apply ratios in SAT questions. For example:

EXAMPLE

A polo club has a 5:7 ratio of members who own at least one horse to members who don't own a horse. If the club has 75 members who own horses, how many members don't own a horse?

Use the ratio 5:7 to build a proportional equation as follows:

$$\frac{5}{7} = \frac{\text{horse owners}}{\text{non-horse owners}}$$

Notice that I've preserved the order of the ratio by placing *horse owners* in the numerator of the fraction and *non-horse owners* in the denominator. Next, plug in information from the problem:

$$\frac{5}{7} = \frac{75}{x}$$

In this case, I've substituted x for *non-horse owners*, because this value is what I'm trying to find. Now, to solve for x, cross-multiply and isolate x by dividing:

$$5x = 525$$
$$\frac{5x}{5} = \frac{525}{5}$$
$$x = 105$$

Therefore, the club has 105 members who don't own a horse, so the correct answer is 105.

This type of problem could be a bit more difficult if the question asks you for a value other than x. For example:

EXAMPLE

A private high school has an 11:2 ratio of in-state to out-of-state students. If the high school has 462 in-state students, what is the total number of students in attendance?

This time, build a proportional equation as follows:

$$\frac{11}{2} = \frac{\text{in-state students}}{\text{out-of-state students}}$$

Plug in the information from the question and solve for x:

$$\frac{11}{2} = \frac{462}{x}$$
$$11x = 924$$
$$x = 84$$

Thus, the school has 84 out-of-state students, and so it has a total of $462 + 84 = 546$ students, so the answer is 546.

Using Ratios and Proportions to Solve SAT Word Problems

When you understand how ratios and proportions work, you can use these tools to solve a variety of SAT word problems. In this section, I show you how to set up proportional equations to answer an assortment of typical SAT questions.

Solving unit conversion problems

One common variety of SAT question asks you to convert one type of unit to another. Typically, the problem provides you with an equivalence for performing the unit conversion. For example:

EXAMPLE

Colorado's southern border is approximately 380 miles in length. On an antique map, one inch represents 75 miles. Which of the following is closest to the length of Colorado's southern border as depicted on the map?

(A) 5.0 inches

(B) 5.1 inches

(C) 5.2 inches

(D) 5.3 inches

When setting up a proportional equation to solve this problem, be sure to include the units. Here, I use the fact that 1 inch on the map equals 75 miles on the left side of the equation, and then let x equal the length I'm looking for:

$$\frac{1 \text{ inch}}{75 \text{ miles}} = \frac{x \text{ inches}}{380 \text{ miles}}$$

When you cross-multiply, all the units can be safely dropped:

$$380 = 75x$$

Divide both sides by 75 to solve the problem:

$$380 = 75x$$
$$\frac{380}{75} = \frac{75x}{75}$$
$$5.07 \approx x$$

Thus, the southern border of Colorado is approximately equal to 5.07 inches, which rounds to 5.1 inches, so Answer B is correct.

Here's another example:

EXAMPLE

The Chinese unit of currency is the yuan, with 1 yuan equal to approximately $0.15 in U.S. currency. When Sherry moved to Nanjing, China, her budget limited her to apartments that cost $400 per month. What was the maximum number of whole yuan that Sherry could afford to pay for an apartment?

With the information that 1 yuan is equivalent to $0.15, let x equal the value you're looking for and set up the following proportional equation:

$$\frac{1 \text{ yuan}}{0.15 \text{ dollar}} = \frac{x \text{ yuan}}{400 \text{ dollars}}$$

Again, with the units being equivalent on both sides of the equation, you can safely drop them as you cross-multiply to remove the fractions:

$$400 = 0.15x$$

To solve for x, divide both sides by 0.15:

$$\frac{400}{0.15} = \frac{0.15x}{0.15}$$
$$2,666.7 \approx x$$

WARNING

Thus, $400 is approximately equal to 2,666.7 yuan. But be careful when answering this question! It asks you to give the number of yuan that is worth *less* than $400. If you have any doubt about this value, multiply both 2,666 and 2,667 by $0.15:

$$2,666 \times \$0.15 = \$399.90 \qquad 2,667 \times \$0.15 = \$400.05$$

Thus, 2,667 yuan is equivalent to slightly more than $400, so the correct answer is 2,666.

Using population density in a proportional equation

Population density is a ratio that's calculated as the number of people in a given area of land. Here's a typical SAT question that asks you to work with population density:

EXAMPLE

New Jersey is the most densely populated U.S. state, with approximately 1,200 people per square mile of land. If the entire state has a population of 9.3 million people, which of the following is the best approximation of New Jersey's total land area?

(A) Between 4,000 and 5,000 square miles

(B) Between 5,000 and 6,000 square miles

(C) Between 6,000 and 7,000 square miles

(D) Between 7,000 and 8,000 square miles

To answer this question, set up the following proportional equation:

$$\frac{1,200 \text{ people}}{1 \text{ square mile}} = \frac{9,300,000 \text{ people}}{x \text{ square miles}}$$

The units are equivalent in the numerator and denominator on both sides of the equation, so when you cross-multiply, you can safely drop these values:

$$1,200x = 9,300,000$$

Divide both sides by 1,200 to complete the problem:

$$\frac{1,200x}{1,200} = \frac{9,300,000}{1,200}$$
$$x = 7,750$$

Therefore, Answer D is correct.

Answering questions involving similar triangles

When a pair of triangles have the same three angles, they're called *similar triangles*. In Chapter 14, I discuss similar triangles along with other types of geometry questions.

Although SAT questions about similar triangles may look like geometry problems, the key to answering them lies in one important fact: The three pairs of corresponding sides of similar triangles are proportional and, therefore, all have the same ratio.

Here's an example that shows you how to work with similar triangles:

EXAMPLE

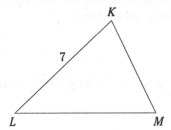 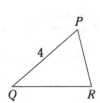

In this figure, $\angle K = \angle P$, $\angle L = \angle Q$, and $\angle M = \angle R$, with corresponding sides in a 7-to-4 ratio as shown. If the length of LM is 5 meters longer than the length of QR, what is the length of QR to the nearest tenth of a meter?

The two triangles have three equivalent angles, so they are similar triangles. Each pair of corresponding sides has the ratio of 7 to 4, so you can use this ratio to set up the following proportional equation:

$$\frac{LM}{QR} = \frac{7}{4}$$

Now, let x equal the length of QR, so $x + 5$ equals the length of LM.

$$\frac{x+5}{x} = \frac{7}{4}$$

To solve for x, begin by cross-multiplying, and then isolate x:

$$4(x+5) = 7x$$
$$4x + 20 = 7x$$
$$20 = 3x$$
$$\frac{20}{3} = \frac{3x}{3}$$
$$6.7 \approx x$$

Therefore, the correct answer is 6.7.

In some cases, the presence of similar triangles in a problem may be a little obscure. Here's another common type of SAT question that hinges on the proportionality of similar triangles:

EXAMPLE

Meghan is 154 centimeters tall, and the height of her younger brother, Jason, is 98 centimeters. She measured Jason's shadow on the sidewalk at 140 centimeters, and then he measured Meghan's shadow. Assuming both measurements were accurate, how many centimeters was Meghan's shadow?

A quick sketch of the problem reveals the two similar right triangles that are the key to answering the question.

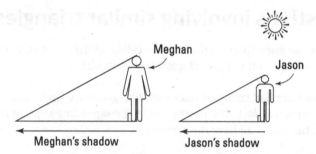

The sides of these two similar triangles are proportional, so each corresponding pair of sides has the same ratio. Thus, you can make the following proportional equation:

$$\frac{\text{Meghan's height}}{\text{Jason's height}} = \frac{\text{Meghan's shadow}}{\text{Jason's shadow}}$$

Fill in the appropriate values into this equation, letting x stand for the length of Meghan's shadow:

$$\frac{154}{98} = \frac{x}{140}$$

To solve for x, begin by cross-multiplying, and then divide both sides by 98:

$$21{,}560 = 98x$$
$$220 = x$$

Therefore, the correct answer is 220.

Understanding direct proportionality

When a pair of values are directly proportional, you can use a proportional equation to find the change in one value based upon how the other value changes. For example:

EXAMPLE

Josephine owns an apartment whose value is directly proportional to the value of the entire building. Five years ago, when she bought the apartment for $155,000, Josephine also noted that the entire building had recently been bought for $2,350,000. This year, the building was sold for 2,600,000. Which of the following is the best estimate of the current value of Josephine's apartment?

(A) $171,500

(B) $172,250

(C) $173,750

(D) $174,500

The value of Josephine's apartment is directly proportional to the value of the building, so you can set up the following proportional equation:

$$\frac{\text{Apartment's previous value}}{\text{Building's previous value}} = \frac{\text{Apartment's current value}}{\text{Building's current value}}$$

Fill in the values given in the problem, letting x stand for the apartment's current value:

$$\frac{155,000}{2,350,000} = \frac{x}{2,600,000}$$

To make this problem a little easier to solve using a calculator, you can simplify the left side by dividing both the numerator and denominator by 1,000:

$$\frac{155}{2,350} = \frac{x}{2,600,000}$$

Solve for x by cross-multiplying and dividing:

$$403,000,000 = 2,350x$$
$$\frac{403,000,000}{2,350} = \frac{2,350x}{2,350}$$
$$171,489.36 \approx x$$

Therefore, the approximate value of Josephine's apartment is \$171,489.36, which is closest to \$171,500, so Answer A is correct.

Using Percentages to Answer SAT Questions

SAT questions that involve percentages can be especially tricky. Many of them are word problems that may seem difficult to make sense of, and even more so to calculate. In this section, I give you a refresher on working with percentages. I also hope to give you a fresh perspective that allows you to answer more difficult percent word problems with relative ease.

Setting up equations to solve percentage problems

Some SAT questions involving percentages are relatively easy to do on your calculator. If you can answer a simple percent question accurately without too much scribbling, go for it.

In most cases on the SAT, however, percentage problems may be confusing. If you're not quite sure how to handle a percentage problem, turn it into an equation and then use algebra — and your calculator, if available — to solve it. For example:

If 37.5% of k is 99, and 132 is p% of k, then what does p equal?

EXAMPLE

To begin, rewrite the statement "37.5% of k is 99" as an equation as follows:

37.5	%	of	k	is	99
37.5	×0.01	×	k	=	99

Here's the finished version of the equation:

$$37.5 \times 0.01 \times k = 99$$

Next, your algebra skills and your calculator can help you solve for k:

$$0.375k = 99$$
$$\frac{0.375k}{0.375} = \frac{99}{0.375}$$
$$k = 264$$

Now, use this value of k to rewrite the statement "132 is p% of k" as follows:

132	is	p	%	of	k
132	=	p	×0.01	×	264

Again, here's the resulting equation:

$$132 = p \times 0.01 \times 264$$

Now, solve for p:

$$132 = 2.64p$$
$$\frac{132}{2.64} = \frac{2.64p}{2.64}$$
$$50 = p$$

Therefore, the correct answer is 50.

Solving percent increase and percent decrease problems

Percent increase and percent decrease problems can be especially confusing. This confusion can be compounded when you're taught to calculate a percentage and then add it to or subtract it from the value that you started with.

In this section, I first refresh you on how you may have been taught to calculate percent increase or decrease. Then, I show you how to calculate percent increase in one step by thinking about it in a different way. This method will also carry over to related problems about exponential increase and decrease.

Finding percent increase and decrease in two steps

Percent increase problems arise, especially in SAT questions, when calculating the following:

» Interest on a loan

» Profit from an investment

» A tip at a restaurant

One way to solve this type of problem is first to calculate the amount of the increase, and then add it to the original value. For example:

EXAMPLE

Monica bought lunch at a diner, and the bill came to $13.50. If she left an 18% tip for the server, what was the total cost of the lunch?

(A) $15.83

(B) $15.93

(C) $16.03

(D) $16.13

This question asks you to calculate an increase of 18%. To answer it in two steps, you can begin by calculating 18% of $13.50:

$$\$13.50 \times 0.18 = \$2.43$$

Now, add this result to $13.50.

$$\$13.50 + \$2.43 = \$15.93$$

Therefore, Answer B is correct.

Percent decrease problems crop up when you're calculating the following:

» The sale price of an item

» Loss from an investment

» Depreciation in value over time

You can answer these types of questions by calculating the amount of the decrease, and then subtracting the result from the original value. For example:

EXAMPLE

Andrea bought a used car for $9,500. She kept it for two years and then sold it for a loss of 27%. To the nearest dollar, what was the selling price of the car?

As with the previous problem, you can calculate this value in two steps. First, multiply $9,500 by 0.27, which is the decimal value of 27%:

$$\$9,500 \times 0.27 = \$2,565$$

Next, subtract this value from the original $9,500 cost of the car:

$$\$9,500 - \$2,565 = \$6,935$$

Therefore, the answer is $6,935, which you can enter into the grid as 6935.

If this method of calculating percent increase and decrease is how you were taught and what works for you, feel free to use it.

However, in the next section, I show you what I consider to be a much better method for solving these types of problems. It's faster, less likely to cause errors, and much easier to apply to harder problems.

Finding the multiplier to calculate percent increase and decrease

With all due respect to the method I show you in the previous section, there's a better way to calculate percent increase and decrease called *finding the multiplier*.

Finding the multiplier allows you to get your answer to a variety of SAT questions in one step using this very handy and powerful formula:

$$\text{Original value} \times \text{Multiplier} = \text{Final value}$$

Best of all, this way of approaching percent increase and decrease is indispensable when you're ready to face down the questions about exponential growth and decay functions awaiting you in Chapter 13.

To find the multiplier, do one of the following:

>> Add percent increase to 100%, and convert the result to a decimal.

>> Subtract percent decrease from 100%, and convert the result to a decimal.

Table 7-1 shows you how to find the multiplier for a variety of percent increase and percent decrease values.

TABLE 7-1 Finding the Multiplier to Calculate Percent Increase and Percent Decrease

Percent Increase			Percent Decrease		
"Percent up"	"Percent of"	Multiplier	"Percent off"	"Percent of"	Multiplier
20%	120% of	1.2	15% off	85% of	0.85
7%	107% of	1.07	6% off	94% of	0.94
2.5%	102.5% of	1.025	3.5% off	96.5% of	0.965

To understand how to apply this formula, take another look at the first question from the previous section:

Monica bought lunch at a diner, and the bill came to $13.50. If she left an 18% tip for the server, what was the total cost of the lunch?

EXAMPLE

(A) $15.83
(B) $15.93
(C) $16.03
(D) $16.13

To answer this question with only one calculation, remember that a percent increase of 18% *up from* a value is equivalent to finding 118% *of* that value, because $100\% + 18\% = 118\%$. So in this case, use a multiplier 1.18 to calculate your answer directly using the formula Original value × Multiplier = Final value:

$$\$13.50 \times 1.18 = \$15.93$$

That's it! Answer B is still correct, but now you've got it in one step.

Here's another take on the second problem from the previous section:

Andrea bought a used car for $9,500. She kept it for two years and then sold it for a loss of 27%. To the nearest dollar, what was the selling price of the car?

EXAMPLE

In this case, notice that 27% *off* a value is the same thing as 73% *of* that value, because $100\% - 27\% = 73\%$. So in this case, use a multiplier of 0.73 with the formula Original value × Multiplier = Final value:

$$\$9,500 \times 0.73 = \$6,935$$

Done! Whether you're working with or without a calculator, one fewer calculation is one fewer chance to make an error.

Identifying percent increase and decrease

In some cases, an SAT question will provide you with a value that changes to a new value and then ask you to calculate the percent increase or decrease reflected in that change.

If you feel comfortable with finding the multiplier, as I show you in the previous section, you can answer this type of question using the following formula:

Original value × Multiplier = Final value

Here's an example:

EXAMPLE

Jared treated a friend to a birthday dinner. When the check arrived, the bill came to $42.35. Because the service was good, he left $50.00 for the server. Which of the following values is closest to the percentage of Jared's tip?

(A) 16%

(B) 17%

(C) 18%

(D) 19%

To find the multiplier for this percent increase, plug the final value of $50 and the original value of $42.35 into the formula Original value × Multiplier = Final value:

$$42.35 \times \text{Multiplier} = 50$$

Divide both sides by 42.35:

$$\text{Multiplier} = \frac{50}{42.35} \approx 1.18$$

A multiplier of 1.18 indicates a percent increase of 18%, so Answer C is correct.

When solving a similar problem involving percent decrease, you may need to subtract the result by 100% to answer certain types of SAT questions. For example:

EXAMPLE

Monica invested $6,700 in a stock and sold it for a loss at $4,800. To the nearest whole percentage, what was the loss on her investment?

(A) 26%

(B) 27%

(C) 28%

(D) 29%

Again, use the formula Original value × Multiplier = Final value, filling in 6,700 for the original value of the stock and 4,800 for the final value:

$$6,700 \times \text{Multiplier} = 4,800$$

Divide both sides by 6,700 to solve for the multiplier:

$$\text{Multiplier} = \frac{4,800}{6,700} \approx 0.716$$

A multiplier of 0.716 indicates a percent decrease of 71.6%. This corresponds to a loss of $100\% - 71.6\% = 28.4\%$, which rounds down to 28%, so Answer C is correct.

Solving tricky reverse percent increase and decrease problems with algebra

The trickiest type of percent increase/decrease problem gives you the final value and asks you to calculate the original value. But when you're comfortable finding the multiplier, you can use the following formula:

Original value × Multiplier = Final value

For example:

EXAMPLE

Sondra made 8.5% profit on an investment in a mutual fund. If she sold the mutual fund for $10,470.25, how much did she pay for it originally?

Here, the problem gives you the final value of $10,470.25 and asks you to calculate the original value. In this case, a percent increase of 8.5% *up from* a value changes to 108.5% *of* that value, so use the multiplier 1.085:

Original value × 1.085 = 10,470.25

Divide both sides by 1.085:

$$\frac{\text{Original value} \times 1.085}{1.085} = \frac{10,470.25}{1.085}$$
$$\text{Original value} = 9,650$$

Therefore, the original price of the index fund was $9,650, which you should grid into the chart as 9650.

Here's a similar percent decrease problem:

EXAMPLE

Cassius found an antique volume of *The Brothers Karamazov* by Dostoevsky for sale online. He negotiated a 12% reduction in the posted price that the seller was asking for, which brought the price down to $677.60. What was the posted price of the book?

In this question, the final value is $677.60 and you're asked to calculate the original value. Here, a percent decrease of 12% *off* the original value translates to 88% *of* the original value, so use a multiplier of 0.88:

Original value × 0.88 = 677.6

Divide both sides by 0.88:

$$\frac{\text{Original value} \times 0.88}{0.88} = \frac{677.6}{0.88}$$
$$\text{Original value} = 770$$

Therefore, the seller's posted price for the book was $770, which you should grid into the chart as 770.

IN THIS CHAPTER

» **Understanding basic statistical calculations, such as** *range, mean, median,* **and** *standard deviation*

» **Spotting flaws in data collection methods**

» **Clarifying how to interpret statistics**

» **Knowing the formula for probability**

» **Calculating compound and conditional probability**

» **Answering SAT probability questions based on two-way tables**

Chapter **8**

Statistics and Probability

n this chapter and the next, you explore how statistics and probability help to make sense of real-world observations.

First, I provide some need-to-know information about statistics, including how to calculate the range, the mean, and the median of a data set. I discuss how an outlier can alter these values in a data set, and give you some general facts about the standard deviation of a data set. Then, I show you how to answer SAT questions that focus on data collection and interpretation of statistical conclusions.

After that, you use the probability formula to calculate probability, including compound and conditional probability. You also apply these methods to answer SAT probability questions based on information provided in two-way tables.

Statistics

Statistics is the mathematical science of collecting and interpreting information about the world. This information usually comes in *data sets* — that is, sets of ordered numerical information based on real-world observation.

When your *collection method* (how you get the data), *statistical calculations* (how you mathematically manipulate the data), and *interpretation* (the conclusions you draw from the data) are all sound, statistical results can be useful for theorizing about the world and predicting the future.

This section gives you a look at some basic statistical concepts, so that you can answer the main types of SAT questions involving statistics.

Statistical calculations

REMEMBER

EXAMPLE

Statistical calculations are mathematical manipulations applied to a *data set* (or *sample set*) — that is, a set of numbers that have been collected through observation of the world. For example:

Nine of Ms. Giverny's advanced chemistry students received the following grades on their most recent lab project:

Abigail	Cait	Douglas	Erin	Joaquin	Kylie	Logan	Travis	Violet
92	84	82	91	89	82	95	79	89

In this section, I'll use this data set for all the examples that follow.

Ordering a data set

REMEMBER

An important first step when working with statistics is to *order* the data set — that is, put the values in order from least to greatest:

79, 82, 82, 84, 89, 89, 91, 92, 95

For the remainder of this section, I refer exclusively to this ordered version of the data set.

Finding the minimum, maximum, and range

REMEMBER

The *minimum* (*min*) and *maximum* (*max*) of a data set are, respectively, the least and greatest values in that set. In this data set, the min is 79 and the max is 95.

The *range* (or *spread*) of a data set is the difference between the min and max. Calculate the range as follows:

$$\text{Range} = \text{Max} - \text{Min} = 95 - 79 = 16$$

Thus, the range for this data set is 16.

Calculating the mean and median of a data set

REMEMBER

The *mean* (or *mean average*) is the center of a data set as calculated in the following way:

$$\text{Mean} = \frac{\text{Sum of values}}{\text{Number of values}}$$

In the example, calculate the mean of the nine chemistry lab grades as follows:

$$\text{Mean} = \frac{79 + 82 + 82 + 84 + 89 + 89 + 91 + 92 + 95}{9} = \frac{783}{9} = 87$$

So the mean of these nine grades is 87.

In contrast, the *median* is the middle value of a data set. If a data set has an even number value, calculate the median as the mean of the two middle values.

Thus, in the example, the median of the nine chemistry lab grades is 89.

Understanding outliers

An *outlier* in a data set is a value that is either much less or much greater than most of the other values in the set.

For example, in the example of Ms. Giverny's chemistry lab grades, suppose a tenth student named Biff has missed the lab. Ms. Giverny allows him to make it up, and his grade is a 37. Thus, the data set now has ten values, as follows:

37, 79, 82, 82, 84, 89, 89, 91, 92, 95

The value 37 is a *low outlier* in the data set. Generally speaking, outliers tend to affect the mean more than the median. To check this fact, I'll recalculate the mean for the data set that now includes this new value:

$$\text{Mean} = \frac{39 + 79 + 82 + 82 + 84 + 89 + 89 + 91 + 92 + 95}{10} = \frac{822}{10} = 82.2$$

Now, I'll recalculate the median as the mean of the two middle values in the new data set:

$$\text{Median} = \frac{84 + 89}{2} = \frac{173}{2} = 86.5$$

As you can see, the presence of a single outlier reduces the mean by 4.8 but reduces the median by only 2.5.

The range is even more greatly affected by the presence of an outlier. In this example, calculate the range of the new data set as follows:

$$\text{Range} = \text{Max} - \text{Min} = 95 - 39 = 56$$

In this case, the range increases from 16 to 56!

Knowing about normal distribution of data sets

When a data set has a *normal distribution*, the values in that set tend to be clustered toward the middle, with lesser and greater values distributed approximately equally. This shape is also described as a *bell-curve distribution*. In a set with a normal distribution, the mean and median are approximately equal.

A good example of a data set with a normal distribution is the set of scores for the SAT Math Test itself. Every year, millions of students take the SAT, and their math scores display a normal distribution in a range of 200 to 800 points, around a mean of approximately 500, as shown in Figure 8-1.

FIGURE 8-1:
The set of
scores for
an SAT Math
Test display
a normal
distribution,
which takes
the shape of
a bell curve.

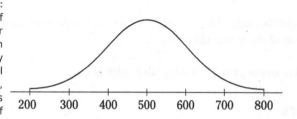

200 300 400 500 600 700 800

Understanding the standard deviation

The *standard deviation* is a measurement of how much the values in a data set differ (or *deviate*) from the mean.

On the SAT, you won't have to perform the time-consuming process of calculating a standard deviation. But you may have to answer questions that demonstrate a general understanding of what the standard deviation tells you about a data set.

REMEMBER

Here are a few rules of thumb for estimating and comparing standard deviations. Generally speaking, the standard deviation tends to increase as:

>> The range of a data set increases.

>> The distribution of values within a data set shift from the center (mean) toward the edges (minimum and maximum values).

EXAMPLE

Data sets M and N each include 20 values, both with a normal distribution. The range of values in data set M equals 18, and the range of values in data set N equals 35. Which of the following statements about the standard deviations of the two data sets is true?

(A) The standard deviation of data set M is less than that of data set N.

(B) The standard deviation of data set M is equal to that of data set N.

(C) The standard deviation of data set M is greater than that of data set N.

(D) There is not enough information to make a determination about standard deviation.

The distribution of both data sets is normal, so there should be no significant difference in the distribution of the sets to affect the standard deviation. Data set M has a much smaller range than data set N, so data set M has a smaller standard deviation. Therefore, Answer A is correct.

Collecting data and interpreting statistics

Some SAT questions don't ask you to calculate statistics from a data set. Instead, a question might give you information about how the data was collected and ask you to identify a flaw in this method that might invalidate the data. Alternatively, a question might ask you to identify a reasonably valid interpretation of a statistical study.

When you know a few important facts about collecting data and interpreting statistics, you can often answer these questions quickly and accurately, sometimes without jotting down a single number. In this section, I show you how to make sense of this type of question.

Understanding sample bias

Statistical information is only as good as the data that it's based on. Another way to say this is, "Garbage in, garbage out."

One way that data can be of questionable value is when it has a *sample bias* — that is, when the data is unrepresentative of the population it purports to model. Here's an example of a question that asks you to consider this concern:

EXAMPLE

The city council of Taborville is considering a proposal to build an outdoor basketball court in a local park. An independent research firm was hired to find information about this proposal's popularity among Taborville residents. They surveyed 235 people and found that only 16% favor the proposal. Which of the following, if true, would LEAST invalidate this finding?

(A) The survey focused mostly on winter vacationers in Taborville staying at an exclusive ski resort nearby.

(B) The survey was placed in bookstores throughout the city, freely available to all patrons who were willing to fill it out.

(C) The survey was collected over a period of less than two months after it was first commissioned.

(D) The survey was conducted at a retirement community located within walking distance of the park where the basketball court would be placed.

Three of these answers include various forms of sample bias, limiting the population participating in the survey to people who might be disinclined to favor the basketball court.

In Answer A, the group surveyed is winter vacationers who come to Taborville to ski and, therefore, might have no interest in an outdoor basketball court. In Answer B, the survey focuses on people in bookstores, who may have less interest in outdoor activities than the general population of Taborville. And in Answer D, the survey is limited to retirees near the park who, presumably, may be less active than the younger population who live nearby.

Only Answer C presents a condition that wouldn't unduly bias the sample of survey participants, so Answer C is correct.

EXAMPLE

A chain of cafes has 107 locations scattered throughout the Midwestern United States. A human resources director who wants to assess employee satisfaction throughout all the locations has developed a survey. Which of the following sampling methods would be most appropriate to this end?

(A) Randomly select one location and survey all employees and customers found at that location on a randomly selected day.

(B) Select the 10 locations with the highest revenue and survey all the employees who work at that location.

(C) Randomly select three employees from each of the 107 locations and survey only those employees.

(D) Distribute surveys to all 107 locations and allow all interested employees to complete the survey.

The director wants to assess employee satisfaction among the group of employees throughout all 107 locations. Answer A is incorrect because selection method limits the survey to only a single location, and incorrectly includes customers in the survey rather than limiting it to employees. Answer B is incorrect because it limits the survey to employees from the locations with the highest revenue. Answer D is incorrect because it allows survey recipients to self-select rather than be randomly selected. Answer C creates a random selection process that includes an equal number of employees from each location, creating a sample set most representative of the employees across all 107 locations, so Answer C is correct.

Keeping an eye on sample size

For statistical conclusions to be valid, a data set should be large enough to reflect the population from which it's drawn. If a sample set is too small in relation to this population to be sufficiently representative, the resulting statistics may be invalid.

Here's an SAT question that focuses on the issue of sample size:

EXAMPLE

Tanya wrote down the names of every student in her graduating high school class on individual slips of paper, placed them in a large bowl, and then drew names at random. For each name she drew, she asked that person to respond to a short survey, and then collected all these surveys when complete. She used this data as the basis of her senior year math project in statistics. Which of the following statements, if true, would provide the LEAST support for the results of Tanya's statistical analysis?

(A) Tanya pulled 5 names from the bowl, and her graduating class has 49 students.

(B) Tanya pulled 27 names from the bowl, and her graduating class has 49 students.

(C) Tanya pulled 5 names from the bowl, and her graduating class has 756 students.

(D) Tanya pulled 49 names from the bowl, and her graduating class has 756 students.

In Answers B and D, Tanya selects a sample set that is, in each case, most likely sufficient to represent the complete population she wants to draw conclusions about. However, a sample size of only 5 people most likely wouldn't provide enough information to draw statistical conclusions. This small sample set would be less problematic if it represented more than 10% of the graduating class, as in Answer A. It would be more problematic if it represented less than 1% of the graduating class, as in Answer C. Therefore, Answer C is correct.

Accounting for margin of error

No statistical analysis is perfect, and any statistical result should provide a *margin of error* — that is, a numerical way to calculate an interval of plausible values for a statistical result. Here's a question that explores the margin of error in more detail:

EXAMPLE

According to a political poll by the Martin Scrubbs for Mayor campaign four weeks before election day, the percentage of voters who currently intend to vote for Scrubbs is 54% with a 3% margin of error. The poll targeted eligible voters inside Scrubbs's district. Which of the following conclusions is most warranted based upon this data?

(A) Scrubbs will win on election day.

(B) No more than 54% of eligible voters currently intend to vote for Scrubbs.

(C) The approximate probability that Scrubbs will win is between 51% and 57%.

(D) Between 51% and 57% of eligible voters currently intend to vote for Scrubbs.

Answer A is incorrect because no statistical information can positively predict a result. The poll targeted eligible voters seeking information about their voting intentions, but isn't intended to calculate the probability that he will win, so Answer C is incorrect. A 54% result with a 3% margin of error results in a plausible conclusion that between 51% and 57% of eligible voters currently intend to vote for Scrubbs, so Answer B is incorrect. Thus, Answer D is correct.

Probability

Probability is the mathematical measurement of how likely a particular outcome of an event is to occur. In this section, I show you how to use the probability formula to answer a variety of SAT probability questions.

Understanding the probability formula

In probability, an *event* is an occurrence that has one or more possible *outcomes*. For example, rolling a single six-sided die is an event that has six possible outcomes, depending upon which side of the die lands face up.

Here's the formula for calculating probability:

$$\text{Probability} = \frac{\text{Target outcomes}}{\text{Total outcomes}}$$

In this formula, the denominator value, *Total outcomes*, is the total number of possible outcomes to a given event. In contrast, the numerator value, *Target outcomes*, is the number of outcomes to an event that satisfy the criteria whose probability you're attempting to calculate.

TIP

When calculating probability, the number of target outcomes is always less than or equal to the number of total outcomes. Start by calculating total outcomes, because this is usually the easier value to calculate. Then think about how many of these outcomes satisfy the conditions stated in the problem, and use this number as your target outcomes.

For example:

EXAMPLE

If you roll a six-sided die, what is the probability that the number 3 will land face up?

In this case, the die can land in six different ways, so the total number of outcomes is 6. However, only one of these outcomes — that the number 3 lands face up — satisfies the conditions of the problem. Thus, you can calculate the probability of this outcome as follows:

$$\text{Probability} = \frac{\text{Target outcomes}}{\text{Total outcomes}} = \frac{1}{6}$$

Therefore, the correct answer is $\frac{1}{6}$, which you can grid in as .166 or .167.

Answering SAT probability questions

In this section, you use the probability formula to answer a variety of common types of SAT probability questions.

Simple probability

REMEMBER

Simple probability is the probability that a single outcome will occur. To calculate simple probability, find the number of total outcomes and target outcomes, and then plug these values into the formula for probability. For example:

EXAMPLE

A bag contains 12 tiles, each inscribed with a different integer, from 1 to 12. If you pull one tile from the bag at random, what is the probability that the number on this tile will be divisible by 3?

The number of total possible outcomes is 12, corresponding to each of the 12 tiles. In this question, the number of target outcomes is 4: the outcomes when you pull out tiles marked 3, 6, 9, or 12. Plug these two values into the formula for probability and simplify:

$$\text{Probability} = \frac{\text{Target outcomes}}{\text{Total outcomes}} = \frac{4}{12} = \frac{1}{3}$$

Therefore, the correct answer is $\frac{1}{3}$, which you can enter into the grid as .333.

Compound probability

REMEMBER

Compound probability is the probability that an outcome meets at least two criteria. To calculate compound probability, find the number of total outcomes, then count the number of outcomes when all the specified criteria are met, and finally plug these values into the formula for probability. For example:

EXAMPLE

A bag contains 12 tiles, each inscribed with a different integer, from 1 to 12. If you pull one tile from the bag at random, what is the probability that the number on this tile will be both divisible by 3 and greater than 5?

As in the previous problem, the number of total possible outcomes is 12. But this time, the number of target outcomes is 3, because only the outcomes when you pull out tiles marked 6, 9, or 12 satisfy both criteria. Plug these two values into the formula for probability and simplify:

$$\text{Probability} = \frac{\text{Target outcomes}}{\text{Total outcomes}} = \frac{3}{12} = \frac{1}{4}$$

Therefore, the correct answer is $\frac{1}{4}$, which also equals .25.

Conditional probability

REMEMBER

Conditional probability is the probability that a target outcome will occur *given that* an initial criterion is met. To calculate conditional probability, first find the number of total outcomes that meet this initial criterion, then count only the outcomes in which this criterion is met *and* the target outcome occurs. Now substitute these two values into values into the formula for probability. For example:

EXAMPLE

A bag contains 12 tiles, each inscribed with a different integer, from 1 to 12. If you pull one tile from the bag at random, what is the probability that the number on this tile will be both divisible by 3 given that this number is greater than 5?

In this question, the initial criterion (that the number must be greater than 5) limits the number of total possible outcomes to 7: when you pull any of the number 6, 7, 8, 9, 10, 11, and 12. Now, consider this pool of 7 numbers, and determine how many of them are also divisible by 3. This value is 3 if you pull the number 6, 9, or 12. Plug these two values into the formula for probability and simplify:

$$\text{Probability} = \frac{\text{Target outcomes}}{\text{Total outcomes}} = \frac{3}{7}$$

Therefore, the correct answer is $\frac{3}{7}$, which you can enter into the grid as either .428 or .429.

Answering table-based SAT probability questions

On the SAT, a common type of probability problem uses information from a *two-way table* — a chart that provides data about a group of individuals.

In this section, I'll use Table 8-1 as the two-way table for all the questions.

TABLE 8-1 **A Two-Way Table for Probability Questions**

	Uses Swimming Pool	Doesn't Use Swimming Pool	Total
Uses Tennis Courts	24	81	105
Doesn't Use Tennis Courts	73	38	111
Total	97	119	216

Finding probability

Here's a question that asks you to calculate simple probability based on the information in Table 8-1:

EXAMPLE

A fitness center surveyed 216 of its members about whether or not they use the swimming pool and the tennis courts. The results are in Table 8-1. What is the probability that a randomly chosen person who completed the survey will be a person who uses the swimming pool?

(A) $\frac{24}{97}$

(B) $\frac{73}{97}$

(C) $\frac{73}{216}$

(D) $\frac{97}{216}$

The total number of people surveyed equals 216. Of these, 97 stated that they use the swimming pool, so this is the target number of outcomes. Place these numbers into the formula for probability:

$$\text{Probability} = \frac{\text{Target outcomes}}{\text{Total outcomes}} = \frac{97}{216}$$

Therefore, Answer D is correct.

Working out compound probability

Compound probability, which I discuss in the previous section, is the probability that an outcome meets at least two criteria. Here's a question about compound probability that utilizes a two-way table:

EXAMPLE

A fitness center surveyed 216 of its members about whether or not they use the swimming pool and the tennis courts. The results are in Table 8-1. What is the probability that a randomly chosen person who completed the survey will be a person who uses the swimming pool but not the tennis courts?

(A) $\frac{73}{97}$

(B) $\frac{97}{111}$

(C) $\frac{73}{216}$

(D) $\frac{111}{216}$

The total number of people surveyed equals 216. Of these, 73 stated that they use the swimming pool but not the tennis courts, so this is the target number of outcomes. Place these two values into the formula for probability:

$$\text{Probability} = \frac{\text{Target outcomes}}{\text{Total outcomes}} = \frac{73}{216}$$

Therefore, Answer C is correct.

Calculating conditional probability

Conditional probability, as I discuss earlier in this chapter, is the probability that a target outcome will occur *given that* an initial criterion is met. Here's a two-way table problem that asks you to calculate conditional probability:

EXAMPLE

A fitness center surveyed 216 of its members about whether or not they use the swimming pool and the tennis courts. The results are in Table 8-1. What is the probability that a randomly chosen person who doesn't use the tennis courts will be among the group that uses the swimming pool?

(A) $\frac{73}{97}$

(B) $\frac{73}{111}$

(C) $\frac{73}{216}$

(D) $\frac{111}{216}$

In this question, the initial criterion is that the person chosen doesn't use the tennis courts, so this limits the total pool of people under consideration to 111. Of these, 73 use the swimming pool, so this is the target number of outcomes. Plug these numbers into the formula for probability:

$$\text{Probability} = \frac{\text{Target outcomes}}{\text{Total outcomes}} = \frac{73}{111}$$

Therefore, Answer B is correct.

» Applying statistics and probability to information presented in tables

» Using qualitative data from bar graphs

» Working with histograms to answer statistics and probability questions about quantitative data

» Reading quantitative data from dot plots

» Understanding how line graphs map information over time

» Knowing how to work with scatterplots

Chapter **9**

Understanding Data and Information from Tables and Graphs

I n Chapter 8, you discovered how to answer a variety of SAT questions focusing on statistics and probability. In this chapter, you use this knowledge to solve problems that present data in a variety of visual forms.

To begin, you work with data in tables. Next, you practice reading qualitative data from bar graphs. Then, you use both histograms and dot plots to understand and work with quantitative data.

After that, I show you how to answer SAT questions that commonly arise from data displayed in line graphs. And to finish up, you work with scatterplots.

Reading Information from Tables

The SAT includes a variety of questions that use tables to convey information. Tables are so ubiquitous, in fact, that you'll find lots of examples of them throughout this book.

Here's an example of a common type of statistics question that uses a table:

EXAMPLE

Jack	Kaitlin	Monroe	Nathan	Teresa	Zach
82	95	89	77	94	

A teacher allowed six students to take a make-up exam for their class. Five of the scores are shown here, but they forgot to record Zach's score. If the mean score for the six tests was 88, what was Zach's score on the test?

One way to answer this question is to use the formula for the mean, as I show you in Chapter 8:

$$\text{Mean} = \frac{\text{Sum of values}}{\text{Number of values}}$$

Plug in 88 for the mean; for the sum of values, plug in the scores you know, using x to represent Zach's score; and finally, for the number of values, use 6:

$$88 = \frac{82 + 95 + 89 + 77 + 94 + x}{6}$$

Now, solve for x:

$$88 = \frac{437 + x}{6}$$
$$528 = 437 + x$$
$$91 = x$$

So the answer is 91.

A second way to solve this type of problem — and a method that some students find a little easier — is to jot down the difference between each student's grade and the average, as follows:

Jack	Kaitlin	Monroe	Nathan	Teresa	Zach
82	95	89	77	94	
–6	+7	+1	–11	+6	

Now, add up the five differences:

$$-6 + 7 + 1 - 11 + 6 = -3$$

Thus, Zach's score must account for +3 added to the average of 88:

$$88 + 3 = 91$$

Therefore, his score is 91.

Understanding Bar Graphs, Histograms, and Dot Plots

Statistical information often comes in visual forms such as bar graphs, histograms, and dot plots.

Bar graphs usually provide *qualitative data* — that is, numerical information about discrete groups that are non-numerical (for example, eye color, occupation, or food preference). In contrast, *histograms* and *dot plots* usually convey *quantitative data* — that is, data about groups that are themselves numerical (for example, height, weight, or age).

In this section, I discuss all these visual ways of displaying data.

Bar graphs

A *bar graph* typically provides qualitative data in a visual way. For example:

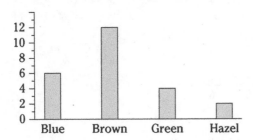

In this section, use this graph to answer each of the questions that follow.

EXAMPLE

This bar graph displays information about the eye color of 24 students in a college statistics class. According to this information, what is the probability that a student chosen at random will have either brown or green eyes?

To answer this question of compound probability, use the probability formula:

$$\text{Probability} = \frac{\text{Target outcomes}}{\text{Total outcomes}}$$

Substitute 24 for the total number of outcomes; then add up the number of students with both brown and green eyes ($12 + 4 = 16$), and substitute this value for target outcomes:

$$= \frac{16}{24} = \frac{2}{3}$$

You can enter this answer into the grid either as the fraction $\frac{2}{3}$ or as the decimals .666 or .667.

Here's another question, this time asking you to calculate conditional probability:

EXAMPLE

According to the information in the bar graph, what is the probability that a student chosen at random who does NOT have green eyes will have hazel eyes?

Again, use the formula for probability.

$$\text{Probability} = \frac{\text{Target outcomes}}{\text{Total outcomes}}$$

In this case, the pool of students who don't have green eyes is $6 + 12 + 2 = 20$, so substitute this value for total outcomes. Next, substitute 2 (the number of students with hazel eyes) for target outcomes:

$$= \frac{2}{20} = \frac{1}{10} = 0.1$$

Therefore, the answer is $\frac{1}{10}$, which you can also enter into the grid as .1.

Histograms

A *histogram* contains numerical information on both axes. This fact can make histograms a bit more challenging to work with than bar graphs, because more complicated calculations are possible.

For example, here's a histogram that provides information about the number of siblings in a population:

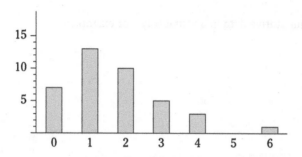

Use this histogram to answer the questions in this section.

EXAMPLE

A group of 39 children at a summer camp were asked for information about the number of siblings that each one has. The histogram shows the results. What is the median number of siblings for the 39 children?

The histogram shows that 7 children have 0 siblings, 13 children have 1 sibling, 10 children have 2 siblings, and so forth. You can arrange this data in order for the 39 children, as follows:

00000001111111111111**1**2222222222233333344446

The middle child in this group has 1 sibling, so the median number of siblings is 1.

Here's a probability question that uses information from this histogram:

EXAMPLE

If a child from this population is chosen randomly, what is the probability that they will have more than 2 siblings?

As usual with probability questions, begin with the formula for probability:

$$\text{Probability} = \frac{\text{Target outcomes}}{\text{Total outcomes}}$$

In this problem, the number of total outcomes when selecting a child at random is 39. And the number of children who have more than 2 siblings is $5 + 3 + 1 = 9$, so this is the number of target outcomes:

$$\frac{9}{39} = \frac{3}{13} \approx 0.2307$$

Therefore, the correct answer is $\frac{3}{13}$, which you can also enter into the grid as .230 or .231.

Dot plots

A *dot plot* is an alternative way to display quantitative data. Like a histogram, it provides a visual way to work with statistical information.

The following dot plot displays the number of goals scored by a soccer team in each of the first 12 games of the season. Use it to answer the questions in this section:

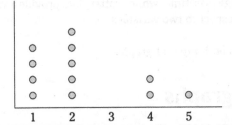

 A soccer team scored from 1 to 5 goals in each of its first 12 games of the season, as shown in the graph. What is the mean number of goals that the team scored per game?

EXAMPLE

To find this answer, notice the information that the dot plot is presenting: The team scored 1 goal in 4 games, 2 goals in 5 games, 4 goals in 2 games, and 5 goals in 1 game. Thus, you can calculate the total number of goals as follows:

$$1(4) + 2(5) + 4(2) + 5(1) = 4 + 10 + 8 + 5 = 27$$

To calculate the mean number of goals per game, divide this value by the number of games played, which is 12:

$$\text{Mean} = \frac{\text{Sum of values}}{\text{Number of values}} = \frac{27}{12} = 2.25$$

Therefore, the team scored a mean number of 2.25 goals per game.

Here's another example that uses this information and takes it further:

 A soccer team scored from 1 to 5 goals in each of its first 12 games of the season, as shown in the graph. What is the difference between the mean number of goals that the team scored per game and the median number of goals that they scored per game?

EXAMPLE

(A) −0.75

(B) −0.25

(C) 0.25

(D) 0.75

To answer this question, you can use the mean number that you calculated in the last problem. To find the median, it may be helpful to arrange the number of goals scored in each of the 12 games in order from least to greatest:

1 1 1 1 2 **2 2** 2 2 4 4 5

Here, the two middle games are in boldface. Because both of these games were two-goal games, the median number of goals scored per game equals 2. To answer the question, subtract the mean minus the median:

$$\text{Mean} - \text{Median} = 2.25 - 2 = 0.25$$

Therefore, Answer C is correct.

Working with Line Graphs and Scatterplots

Two additional types of graphs that are loosely based on the *xy*-plane are called line graphs and scatterplots. Line graphs usually depict change over time, while scatterplots provide a visual representation of a population displayed with respect to two variables.

In this section, I show you how to work with both types of graphs.

Mapping time with line graphs

Line graphs almost always display how change occurs over time, with the horizontal axis most often representing the time dimension. For example:

This line graph shows the percentage of water in a swimming pool during the 16-hour process of draining it, cleaning it, and refilling it.

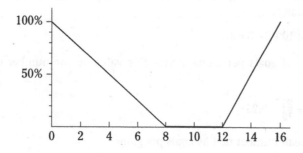

Here are a couple of SAT questions based on this line graph.

EXAMPLE

According to the graph, how many hours longer did the pool take to drain than to refill?

The graph shows that the pool took 8 hours to drain and 4 hours to fill, so it took $8 - 4 = 4$ hours longer to drain than to fill, and so the correct answer is 4.

EXAMPLE

After the pool was cleaned, approximately how many <u>minutes</u> did it take to refill to 80% capacity?

The graph shows that the pool took 4 hours to refill, which equals 240 minutes. Calculate 80% of this value as follows:

$$240 \times 0.8 = 192$$

Therefore, the pool took 192 minutes to refill to 80% capacity.

Understanding scatterplots

A *scatterplot* is a graph containing points representing pairings of numeric values. A scatterplot provides a visual representation of how two variables correspond with each other.

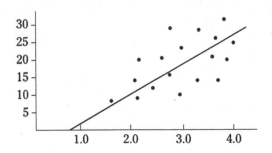

For example, this scatterplot shows data points representing 18 different high school students. The horizontal axis shows each student's grade point average, and the vertical axis shows the number of hours that each student reports studying outside of class each week.

The diagonal line drawn through this graph, called the *line of best fit*, represents the best numerical relationship between the two variables of grade point average and study time.

Use this graph to answer the questions that follow.

EXAMPLE

Which of the following equations best approximates the line of best fit for the scatterplot, with x as the input variable for grade point average and y as the output variable for weekly hours of study time?

(A) $y = 6x + 8$

(B) $y = 6x - 8$

(C) $y = 8x + 6$

(D) $y = 8x - 6$

The line of best fit crosses the x-axis, so it would have a negative y-intercept, which rules out Answers A and C. The line also appears to intersect the point $(1,2)$, so try plugging these values into the other two equations to see which works best:

$y = 6x - 8$ $y = 8x - 6$

$2 = 6(1) - 8$ WRONG! $2 = 8(1) - 6$

$2 = 6 - 8$ $2 = 8 - 6$

$2 = -2$ $2 = 2$

This point works for Answer D, so that is the correct answer.

EXAMPLE

Meghan is the student represented on the graph who has the highest grade point average of all 18 classmates. According to the graph, approximately how many hours per week does she study?

(A) 15

(B) 20

(C) 25

(D) 30

The data point representing Meghan is furthest to the right, just above 4.0 on the horizontal axis. This data point appears to be equivalent to 25 on the vertical axis, so Meghan spends about 25 hours a week studying. Therefore, Answer C is correct.

4

Your Passport to Advanced Math

IN THIS CHAPTER

» **Making sense of function notation**

» **Understanding combinations and compositions of functions**

» **Finding the inverse of a function**

» **Knowing some common parent functions**

» **Working with the most important transformations of functions**

Chapter **10**

Functions

Many students want to run away whenever they see the notation $f(x)$. Unfortunately, functions are an important topic in algebra, and they loom large on the SAT, so this response isn't your best option.

The good news is that functions aren't nearly as scary as you may think they are. In this chapter, I begin at the very beginning, showing you how function notation is just a compact way to allow you to plug a value into an expression.

After some practice working with a few basic functions in this way, you move up to adding, subtracting, multiplying, and dividing functions. Then, you discover how to work with compositions of functions (functions within functions). You also learn how to find the inverse of a function when one exists.

Next, you graduate to graphing functions on the xy-plane. You use the vertical line test to decide whether a graph is a function. Then, I show you ten important parent functions — four polynomial functions and six other basic functions that you'll need to know for the SAT. Finally, I present three transformations of functions that allow you to move or otherwise change the graph of a function.

Understanding Function Notation

Function notation — such as $f(x)$ or $g(x)$ — can look confusing, but it's really just a quick way to plug a number into an expression. In this section, I do my best to demystify this important concept.

Throughout this section, I use the following two functions for all the examples:

$$f(x) = x^2 + 1 \qquad g(x) = 3x - 4$$

Evaluating a function by plugging in a number

Function notation is just shorthand for plugging a number into an expression — that is, substituting a number for a variable. (If you need a quick refresher on evaluating expressions, flip to Chapter 3.)

For example, if I tell you that $x = 3$, and then ask you to plug that value into the expression $x^2 + 1$, you'll probably tell me that the result is 10, because:

$$3^2 + 1 = 9 + 1 = 10$$

Using the function $f(x) = x^2 + 1$, you can express this same idea more compactly as:

$$f(3) = 10$$

In a similar way, you can declare the function $g(x) = 3x - 4$, and then evaluate $g(5)$ as 11, because:

$$g(5) = 3(5) - 4 = 15 - 4 = 11$$

Not too horrible, right?

Evaluating a function plus or minus a value

You can add a number to a function by simply tacking that value onto the end of it. For example:

EXAMPLE

$$f(x) = x^2 + 1$$

In this function, what is the value of $f(4) + 3$?

To begin, plug 4 into the expression $x^2 + 1$, and add 3 at the end:

$$= 4^2 + 1 + 3$$

Now, just crunch the numbers, using trusty PEMDAS:

$$= 16 + 1 + 3 = 20$$

Therefore, the correct answer is 20.

You can subtract a number from a function in a similar way.

EXAMPLE

Given the function $g(x) = 3x - 4$, what is the value of $g(10) - 5$?

This time, plug 10 for x into $g(x) = 3x - 4$, and subtract 5:

$$= 3(10) - 4 - 5$$

Again, the result isn't too difficult to calculate:

$$= 30 - 4 - 5 = 21$$

Therefore, the correct answer is 21.

Evaluating a function times or divided by a value

To multiply a function by a number, attach that number to the front of the function, just as you would do with a variable. For example:

EXAMPLE

What is the value of $3f(2)$ if $f(x) = x^2 + 1$?

To make sense of the notation $3f(2)$, plug in 2 for x into $f(x)$ as usual, and then use parentheses to multiply the whole thing by 3:

$$3f(2) = 3(2^2 + 1)$$

The parentheses on the right side of the equation are important, because they tell you to evaluate what's inside them *before* you multiply by 3:

$$= 3(4 + 1) = 3(5) = 15$$

Therefore, the correct answer is 15.

In a similar way, you can apply division to function notation. For example:

$$g(x) = 3x - 4$$

EXAMPLE

Using this function, what is $\frac{g(6)}{2}$?

Here, begin by substituting 6 for x into $g(x)$:

$$\frac{g(6)}{2} = \frac{3(6) - 4}{2}$$

Again, when you've finished plugging in the numbers, the math isn't hard to work out:

$$= \frac{18 - 4}{2} = \frac{14}{2} = 7$$

Therefore, the correct answer is 7.

Here's one final example that puts together a lot of what I've discussed so far:

$$f(x) = x^2 + 1$$

EXAMPLE

What is the value of $\frac{-2f(6) - 3}{7} + 12$, given this function?

Your first step is simply to switch out the function notation by plugging in 6 for x into the function $f(x) = x^2 + 1$:

$$\frac{-2f(6) - 3}{7} + 12 = \frac{-2(6^2 + 1) - 3}{7} + 12$$

Before you continue, make sure that you understand how this substitution works. Now, the rest is just PEMDAS. Begin by evaluating the expression inside the parentheses:

$$= \frac{-2(36 - 1) - 3}{7} + 12 = \frac{-2(37) - 3}{7} + 12$$

Now, take care of the multiplication and subtraction in the numerator, then handle the fraction, and finally the addition:

$$= \frac{-77}{7} + 12 = -11 + 12 = 1$$

Therefore, the correct answer is 1.

Evaluating a function by plugging in a variable or an expression

So far, you've substituted numerical values into functions, but you can also substitute in a variable like n or an expression like $k - 2$ or $5p + 3$. Some students find this process initially confusing, but end up finding it easier than plugging in a number because there's usually less number crunching to do.

Substituting a simple variable is probably the easiest thing you can do with a function. For example, again using the function $f(x) = x^2 + 1$, here's how you evaluate $f(n)$:

$$f(n) = n^2 + 1$$

That's it! After you make the variable substitution, there's literally no work left to do.

Plugging a variable expression into a function is not much more difficult when you know what to do. For example:

$$g(x) = 3x - 4$$

EXAMPLE What is $g(k - 2)$, given this function?

(A) $3k - 2$

(B) $3k - 6$

(C) $3k - 10$

(D) $3k - 14$

Plug in $k - 2$ for x into the function:

$$g(k - 2) = 3(k - 2) - 4$$

Now, simplify this expression by distributing the 3 and combining like terms — nothing you can't handle:

$$3k - 6 - 4 = 3k - 10$$

This result cannot be simplified further, so Answer C is correct.

In the following example, I mix in some of the skills you've already learned:

$$f(x) = x^2 + 1$$

EXAMPLE Given this function, what is the value of $2f(5p + 3) - 4$ in terms of p?

(A) $50p^2 + 60p + 12$

(B) $50p^2 + 60p + 16$

(C) $50p^2 + 60p + 20$

(D) $50p^2 + 60p + 24$

To begin, substitute $5p + 3$ for x into the function $f(x) = x^2 + 1$:

$$= 2[(5p + 3)^2 + 1] - 4$$

Before moving on, notice that the expression inside the brackets, $(5p + 3)^2 + 1$, is a direct substitution into the function. This sub-expression is then multiplied by 2, and then 4 is subtracted from the result.

Now, follow the PEMDAS rules, evaluating outward from the parentheses, starting with the exponent, and simplifying as you go:

$$= 2\left[(5p+3)(5p+3)+1\right]-4$$
$$= 2\left[25p^2+15p+15p+9+1\right]-4$$
$$= 2\left[25p^2+30p+10\right]-4$$

At this point, all that's left to do is to distribute and combine like terms:

$$= 50p^2+60p+16$$

Therefore, Answer B is correct. The good news here is that this example is more complicated than a function notation question on your actual SAT is likely to be.

Working with Function Notation

When you understand the basic rules of function notation, you're ready to begin working with functions at a higher level. In this section, you discover how to combine functions using the basic four operations, how to create compositions of functions, and how to find the inverse of a function.

Once again, in this section, I use the following two functions for all your functional needs:

$$f(x) = x^2 + 1 \qquad g(x) = 3x - 4$$

Combining functions

You can combine functions using two different types of notation — standard notation and special notation.

Standard notation is the most common on the SAT. It's easy to read because it uses the symbols and conventions that you're already used to, such as + and −. It also allows you to input different numbers into different functions.

However, the special notation for combining functions may creep into SAT questions in the future. It's not difficult, so I'm including it here, especially for students who are queued up to score 700 or higher.

Combining functions using standard notation

You can combine a pair of functions using addition, subtraction, multiplication, or division. The SAT makers love these types of problems because they look intimidating, but they're usually fairly easy to solve. For example:

EXAMPLE

Given the functions $f(x) = x^2 + 1$ and $g(x) = 3x - 4$, what is the value of $f(7) + g(5)$?

To evaluate this expression, substitute 7 into f and 5 into g, respectively:

$$= (7^2 + 1) + [3(5) - 4]$$

Although some of the parentheses and brackets aren't strictly necessary here, it's better to use them and not need them than to need them and not use them. Now, use PEMDAS to complete the problem:

$$= (49+1)+(15-4)=50+11=61$$

Therefore, the correct answer is 61.

Here's an example that involves subtraction:

$$f(x)=x^2+1 \qquad g(x)=3x-4$$

EXAMPLE Using these two functions, what does $g(5)-f(-2)$ equal?

Again, begin by substituting the given values into the respective functions (and don't miss the fact that here, g comes before f!):

$$=[3(5)-4]-[(-2)^2+1]$$

This time, you really do need the second set of brackets to make sure you subtract everything that is inside them:

$$=(15-4)-(4+1)=11-5=6$$

Therefore, the correct answer is 6.

You can also multiply a pair of functions. For example:

EXAMPLE What is the value of $f(4)\cdot g(3)$ if $f(x)=x^2+1$ and $g(x)=3x-4$?

Just as in the previous examples, plug each value into its respective function:

$$=(4^2+1)[3(3)-4]$$

Note that in this case, I separate the two functions using parentheses and brackets without an operator between them, indicating that the two values inside are to be multiplied. Use PEMDAS to evaluate from the inside and work your way out:

$$=(16+1)(9-4)=(17)(5)=85$$

Therefore, the correct answer is 85.

To divide a pair of functions, use the fraction bar. For example:

$$f(x)=x^2+1 \qquad g(x)=3x-4$$

EXAMPLE Given these functions, what is the value of $\dfrac{f(7)}{g(2)}$?

Begin by plugging in the proper values, and evaluate:

$$=\frac{7^2+1}{3(2)-4}=\frac{49+1}{6-4}=\frac{50}{2}=25$$

Therefore, the correct answer is 25.

Combining functions using special notation

In Table 10-1, I include special notation for adding, subtracting, multiplying, and dividing functions. This notation is limited in that it only allows you to combine two functions using the same input value x.

TABLE 10-1 ## Notation for Combining Functions

Special Notation	Standard Notation
$(f+g)(x)$	$f(x)+g(x)$
$(f-g)(x)$	$f(x)-g(x)$
$(fg)(x)$	$f(x)g(x)$
$(\frac{f}{g})(x)$	$\frac{f(x)}{g(x)}$

Here's an example:

$$f(x) = x^2 + 1 \qquad g(x) = 3x - 4$$

EXAMPLE What is the value of $(\frac{f}{g})(3)$, given these two functions?

To begin, change the special notation to the function notation you've grown to know and love:

$$(\frac{f}{g})(3) = \frac{f(3)}{g(3)}$$

Notice that in making this change, I substituted the value 3 into both functions. From here, simply convert the function notation for f and g as usual:

$$= \frac{3^2 + 1}{3(3) - 4} = \frac{10}{5} = 2$$

Therefore, the correct answer is 2.

This notation for combining functions isn't currently on the SAT, but if that changes, you'll be ready.

Compositions of functions

When you apply a function to a function, the result is called a *composition of functions*. Compositions of functions may look confusing at first, but the key is to evaluate the inner function first, and then evaluate the outer function.

An example should help clarify this process. Consider this question, which involves the composition of functions:

$$f(x) = x^2 + 1 \qquad g(x) = 3x - 4$$

EXAMPLE

Given these two functions, what is the value of $f(g(5))$?

Note that this notation has two sets of nested parentheses, which means that the inside expression gets evaluated before the outside expression. So to begin, you want to find the value of $g(5)$:

$$g(5) = 3(5) - 4 = 11$$

Thus, you can substitute 11 for $g(5)$ into the original problem:

$$f(g(5)) = f(11)$$

At this point, you simply need to find the value of $f(11)$ to get your answer.

$$= 11^2 + 1 = 122$$

Therefore, the correct answer is 122.

Just as when combining functions, composition of functions has its own arcane little notation:

$$(f \circ g)(x) = f(g(x))$$

Although this notation isn't currently used on the SAT, if that changes, you might find it helpful to know how to use it. Consider the following problem:

$$f(x) = x^2 + 1 \qquad g(x) = 3x - 4$$

What does $(g \circ f)(4)$ equal, given these two functions?

Notice that in this problem, g comes before f, and order usually makes a big difference when composing functions. Rewrite this notation with a slightly less perplexing notation:

$$= g(f(4))$$

As in the previous example, start at the center, evaluating $f(4)$ as follows:

$$f(4) = 4^2 + 1 = 17$$

Plug in 17 for $f(4)$ into the composition of functions and then solve:

$$g(f(4)) = g(17) = 3(17) - 4 = 47$$

Therefore, $g(f(4)) = 17$, so the correct answer is 17.

Inverse functions

A pair of *inverse functions* — $f(x)$ and $f^{-1}(x)$ — is similar to a pair of inverse operations, such as adding and subtracting. When you add a value to a number and then subtract that value from the result, you get the number you started with.

Similarly, when you apply a function to a number and then apply the inverse function to the result, the process brings you back to the number you started with. That is:

$$f^{-1}(f(x)) = x$$

Assuming an inverse function exists, you can find the inverse $f^{-1}(x)$ for any function $f(x)$ by using the following steps.

REMEMBER

1. **Change the function notation to y.**

2. **Rewrite every x as y, and every y as x.**

3. **Solve the result for y.**

4. **Change the y to inverse function notation.**

For example, consider the function $g(x) = 3x - 4$.

REMEMBER

1. **Change the function notation to y.** In this example, $g(x) = 3x - 4$ becomes:

   $y = 3x - 4$

2. **Rewrite every x as y, and every y as x.** This changes the equation to:

   $x = 3y - 4$

3. **Solve the result for y.** Begin by adding 4 to both sides, and then dividing both sides by 3:

   $x + 4 = 3y$

   $\frac{x+4}{3} = y$

4. **Change the y to inverse function notation.**

   $g^{-1}(x) = \frac{x+4}{3}$

Thus, $g(x) = 3x - 4$ and $g^{-1}(x) = \frac{x+4}{3}$ are a pair of inverse functions. To show that this procedure works, you can test it by composing those functions, and then substituting $3x - 4$ for $g(x)$:

$$g^{-1}(g(x)) = g^{-1}(3x - 4)$$

Next, plug in $3x - 4$ for x into the expression $\frac{x+4}{3}$ and simplify:

$$= \frac{3x - 4 + 4}{3} = \frac{3x}{3} = x$$

Thus, $g^{-1}(g(x)) = x$, so the functions are an inverse pair.

Using the Vertical Line Test for Functions

When you're working on the xy-plane, you can use the *vertical line test* to determine whether a graph is a function:

REMEMBER

The vertical line test states that a graph is a function if and only if no vertical line passes through the graph more than once.

Understanding this relatively simple idea can be a great way to pick up a quick point or two on your SAT. For example:

EXAMPLE

Which of the following graphs is a function?

(A)

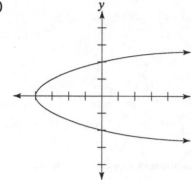

(B)

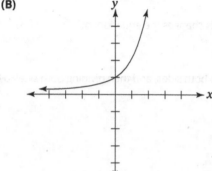

(C)

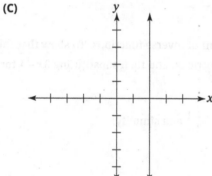

(D)

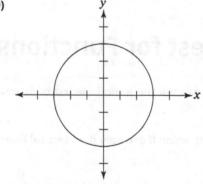

Answers A, C, and D all show graphs that fail the vertical line test. That is, you can pass at least one vertical line through each of these graphs that passes through more than one point on the graph. Thus, all these answers are wrong.

In contrast, the graph shown in Answer B passes the vertical line test: No matter where you draw a vertical line on this graph, it passes through no more than one point. Therefore, Answer B is correct.

Knowing Some Key Parent Functions

A *parent function* is the most basic version of a particular type of function.

Every parent function has a signature shape. Variations of a parent function that preserve its fundamental shape are called — wait for it — *child functions* of that parent function.

Every parent function also passes through a key point on the graph called its *anchor point* or *locator point*.

Knowing the shape and anchor point of a function can help you sketch it from memory when an SAT question doesn't provide you with a graph. This information can also help you to identify the function when a question gives you a graph without an equation.

In this section, I give you an overview of the ten most common families of functions on the SAT. Then, in the next section, I show you how to apply the most common transformations to parent functions.

Polynomial functions

Technically speaking, a *polynomial* is any function of the following form:

$$f(x) = a_n x^n + a_{n-1} x^{n-1} + a_{n-2} x^{n-2} \ldots + a_2 x^2 + a_1 x + a_0$$

I'll admit that's not very helpful to most students. A more useful way to understand polynomials is to realize that most of the functions you deal with in algebra — such as linear and quadratic functions — are varieties of polynomials.

Take a look at Table 10-2 and Figure 10-1 to familiarize yourself with the four types of polynomials you're most likely to see on the SAT. Notice that although each of these types has a different signature shape, they all have an anchor point at (0,0).

In this section, I discuss these four functions in a way that sheds light on what they all have in common. Then, in Chapter 11, I go into a lot more detail, showing you what you need to know about polynomials to do well on the SAT.

TABLE 10-2 **Parent Functions for the First Four Polynomials**

Degree	Name	Function	Anchor Point
1	Linear	$f(x) = x$	(0,0)
2	Quadratic	$f(x) = x^2$	(0,0)
3	Cubic	$f(x) = x^3$	(0,0)
4	Quartic	$f(x) = x^4$	(0,0)

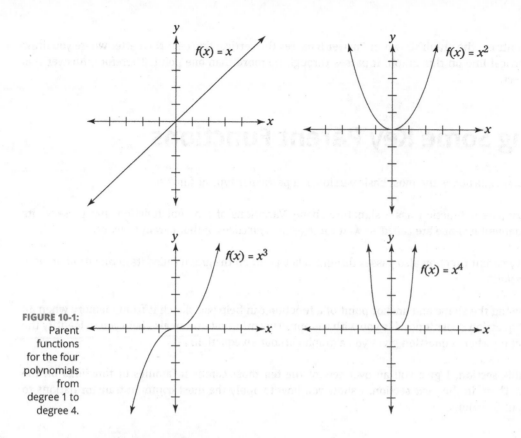

FIGURE 10-1:
Parent functions for the four polynomials from degree 1 to degree 4.

Linear functions

The parent function for the linear polynomial is $f(x) = x$. This graph shows up as a diagonal line that passes through the origin with a slope of 1, as shown in Figure 10-1.

All other linear functions are variations of this parent function, and also show up as straight lines on the xy-plane.

I discuss linear functions in more depth in Chapter 5.

Quadratic functions

The parent function for the quadratic polynomial is $f(x) = x^2$. This is graphed as a parabola — that is, a U-shaped graph — whose vertex (lowest point) is at the origin, as shown in Figure 10-1.

All other quadratic functions are variations of this parent function, and are also parabolas that are either U-shaped or inverted U-shaped graphs.

In Chapter 12, I discuss quadratic functions in greater detail.

Cubic functions

The parent function for the cubic polynomial is $f(x) = x^3$. This graph appears on the xy-plane as a bent curve that crosses the origin, as shown in Figure 10-1.

All other cubic functions are variations of this parent function. In some cases, more complex cubic functions have two bends in them, and resemble the drain pipe under your kitchen sink.

Quartic functions

The parent function for the quartic polynomial is $f(x) = x^4$. This graph resembles the U-shaped quadratic function, but is flatter on the bottom, as shown in Figure 10-1.

All other quartic functions are variations of this parent function. More complex quartic functions can have as many as three bends in them, and tend to look like curvy versions of either the letter W or the letter M.

Other important parent functions

Beyond the polynomial functions, which I discuss in the previous section, there are six additional functions that you're likely to see on the SAT. These are listed in Table 10-3, and also graphed in Figure 10-2.

TABLE 10-3 ## Six Important Parent Functions

Name	Function	Anchor Point
Absolute value	$f(x) = |x|$	(0,0)
Exponential	$f(x) = 2^x$	(0,1)
Radical	$f(x) = \sqrt{x}$	(0,0)
Rational	$f(x) = \dfrac{1}{x}$	(1,1)
Sine	$f(x) = \sin x$	(0,0)
Cosine	$f(x) = \cos x$	(0,1)

In this section, I give you a brief introduction to each of these parent functions.

Absolute value functions

The parent function for the absolute value function is $f(x) = |x|$. This results in the V-shaped graph shown in Figure 10-2. Note that the anchor point for the parent absolute value function is (0,0). As with the quadratic function, this point is also called the vertex of the function.

Exponential functions

Technically speaking, the parent exponential function is $f(x) = e^x$. However, the value e as the base of an exponential function lies outside the current scope of the SAT.

Instead, you can use 2 as a good stand-in for e as the base of an exponential function. Thus, here's a reasonable alternative version of the parent exponential function:

$$f(x) = 2^x$$

This function passes through the anchor point (0,1). As you can see in Figure 10-2, as x increases in the positive direction, this function increases infinitely. However, as x decreases in the negative direction, it gets closer and closer to the x-axis — that is, it has an asymptote at $x = 0$ in the negative direction.

I discuss exponential functions further in Chapter 13.

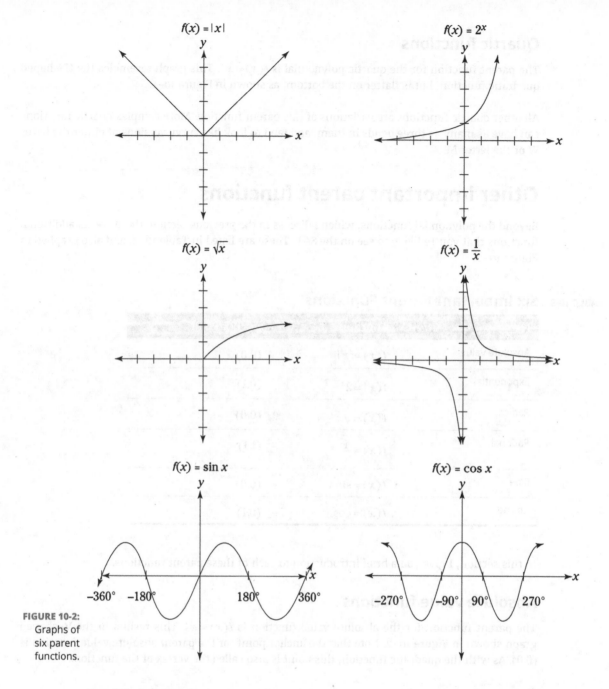

$f(x) = |x|$

$f(x) = 2^x$

$f(x) = \sqrt{x}$

$f(x) = \frac{1}{x}$

$f(x) = \sin x$

$-360°$ $-180°$ $180°$ $360°$

$f(x) = \cos x$

$-270°$ $-90°$ $90°$ $270°$

FIGURE 10-2:
Graphs of
six parent
functions.

Radical functions

The parent function for the radical function is $f(x) = \sqrt{x}$. As Figure 10-2 shows you, this results in a graph that starts at the anchor point (0,0) and slowly rises as x increases.

Because a radical cannot accept negative values as inputs, the domain of the radical function is $x \geq 0$.

You can discover more about radical functions in Chapter 13.

Rational functions

The parent function for the rational function is $f(x) = \frac{1}{x}$. As shown in Figure 10-2, this function has an anchor point at $(1,1)$. It also has a vertical asymptote on the y-axis and a horizontal asymptote on the x-axis.

Rational functions are, for the most part, outside the scope of the SAT. Even so, you may find knowing this parent function handy for answering a more advanced question, so keep it in mind if your SAT math goal is 700 or higher.

Trigonometric functions

Although there are six trigonometric functions, two of these are by a massively wide margin the most commonly used, both on the SAT and elsewhere: the sine and cosine functions.

These functions produce wave-shaped graphs that are simple to sketch and useful for modeling all sorts of physical phenomena, from light and electricity to sound and ocean waves.

Both of these functions are *periodic functions*, which means that they repeat at regular intervals. The period of the sine and cosine parent functions is 360°, which equals 2π radians. So, when you know how to sketch the sine or cosine graph from 0° to 360°, you have a complete picture of the graph.

In this section, I give you the most basic information you need for sketching sine and cosine curves. For a more complete discussion of trigonometry in general, flip to Chapter 14.

SINE FUNCTION

The anchor point of the sine function is $(0,0)$. As you can see from Figure 10-2, the sine curve:

>> At 0° starts at 0

>> At 90° rises to 1

>> At 180° falls back to 0

>> At 270° falls further to –1

>> At 360° returns to 0

Knowing these five points can help you make a quick sketch of the sine curve.

COSINE FUNCTION

The anchor point of the cosine function is $(0,1)$. As shown in Figure 10-2, the cosine curve:

>> At 0° starts at 1

>> At 90° falls to 0

>> At 180° falls further to –1

>> At 270° rises to 0

>> At 360° returns to 1

When you know these five points on the cosine graph, you can graph the function easily.

Transforming Functions

When you know a few parent functions, as I introduce in the previous section, you're ready to understand *transformations of functions* — that is, numerical changes to functions that move or otherwise alter their graphs in predictable ways.

In your current math class, you may work on a variety of transformations that change the graph in all sorts of ways, requiring some very sophisticated calculations. On the SAT, you can primarily focus on these three transformations:

>> Vertical transformations: $f(x) + k$

>> Horizontal transformations: $f(x - h)$

>> Stretch/compress/reflect transformations: $af(x)$

Vertical and horizontal transformations move a function without changing its basic shape. In contrast, a stretch/compress/reflect transformation changes the shape of a function while keeping it essentially in place.

In this section, you work with these three types of transformations, applying them to a variety of parent functions.

Clarifying vertical and horizontal transformations

Vertical and horizontal transformations move a function to a different position on the xy-graph without changing its basic shape on the graph.

In this section, I show you how to work with these two most basic types of transformations both individually and together.

Vertical transformations

A *vertical transformation* of a function moves the entire function either up or down on the xy-graph without affecting its basic shape.

The vertical transformation $f(x) + k$ moves the function $f(x)$:

>> **Up k units when the sign is positive:** For example, $f(x) + 2$ moves the function $f(x)$ *up 2 units*.

>> **Down k units when the sign is negative:** For example, $f(x) - 3$ moves the function $f(x)$ *down 3 units*.

This transformation occurs *outside* the parentheses. Thus, it simply adds a value to or subtracts a value from the entire function, moving it either up or down.

Figure 10-3 shows transformations of three functions that you know from earlier in this chapter. Note in each case that the value k moves the anchor point either up or down from its usual position in the function.

Here's an example of an SAT math question that you can answer when you understand vertical transformations:

$f(x) = 2^x + 1$

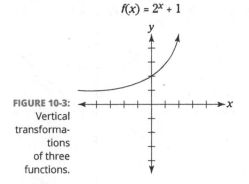

$f(x) = x^2 - 4$

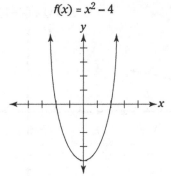

$f(x) = x^3 - 1$

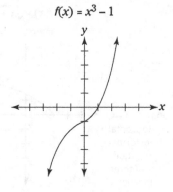

FIGURE 10-3:
Vertical
transforma-
tions
of three
functions.

Which of the following transformations moves the vertex of the function $f(x) = x^2$ to the point $(0,3)$?

EXAMPLE

(A) $f(x) = x^2 + 3$

(B) $f(x) = x^2 - 3$

(C) $f(x) = (x + 3)^2$

(D) $f(x) = (x - 3)^2$

Recall from the previous section that the vertex $(0,0)$ is the anchor point for the quadratic function $f(x) = x^2$. Moving the vertex to $(0,3)$ is a transformation that moves the graph up 3 units. Thus, this transformation is $f(x) = x^2 + 3$, so Answer A is correct.

Horizontal transformations

A *horizontal transformation* of a function moves the entire function either left or right on the *xy*-graph without affecting its basic shape.

The horizontal transformation $f(x - h)$ can be a little tricky, because the minus sign moves the function $f(x)$ *opposite* from the direction that makes intuitive sense:

>> **Right *h* units when the sign is negative:** For example, when $h = 4$, the transformation becomes $f(x - 4)$, which moves the function $f(x)$ *right 4 units*.

>> **Left *h* units when the sign is positive:** For example, when $h = -5$, the transformation becomes $f(x + 5)$, which moves the function $f(x)$ *left 5 units*.

This transformation occurs *inside* the parentheses. Thus, it either adds a value to or subtracts a value from the variable *before* the function is evaluated.

The big takeaway here is that the horizontal transformation moves the function in the *opposite* direction from the one that makes intuitive sense. For example, the transformation $f(x + 1)$ moves the function one unit to the left — that is, in the *negative* direction. And the transformation $f(x - 1)$ moves the function 1 unit to the right — that is, in the *positive* direction.

WARNING

Figure 10-4 shows transformations of three functions that you're already familiar with from earlier in this chapter.

Here's an example of an SAT math question that draws from your knowledge of horizontal transformations:

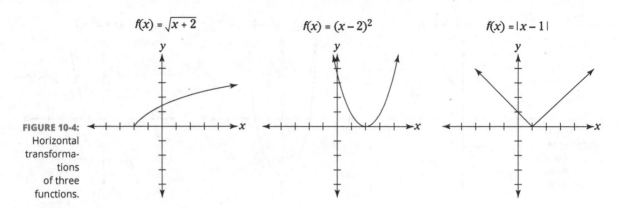

$f(x) = \sqrt{x+2}$ $f(x) = (x-2)^2$ $f(x) = |x-1|$

FIGURE 10-4:
Horizontal
transforma-
tions
of three
functions.

EXAMPLE

Which of the following transformations moves the function $f(x) = x^3$ one unit to the right?

(A) $f(x) = x^3 + 1$

(B) $f(x) = x^3 - 1$

(C) $f(x) = (x+1)^3$

(D) $f(x) = (x-1)^3$

Moving a function one unit to the right is a horizontal transformation, so it occurs inside the parentheses. Thus, you can rule out Answers A and B. This transformation is in the *positive* direction, so it requires a *negative* value for *h*. Therefore, Answer D is correct.

Distinguishing vertical and horizontal transformations

Most students learn about vertical and horizontal transformations around the same time. As a result, these two transformations can be a little confusing. So let me take a few moments of your time to help you make sense of how these two transformations are similar and how they are different.

Table 10-4 gives you some of the key distinguishing features of vertical and horizontal transformations.

TABLE 10-4   Distinguishing Vertical and Horizontal Transformations

Vertical Transformations $f(x) + k$	Horizontal Transformations $f(x - h)$
Move the function up or down.	Move the function left or right.
Constant (*k*) occurs *outside* the function.	Constant (*h*) occurs *inside* the function.
Are *intuitive*. They move the graph in the direction that makes sense:	Are *counterintuitive*. They move the graph in the opposite direction from what makes sense:
In the *positive* direction when the sign is positive	In the *negative* direction when the sign is positive
In the *negative* direction when the sign is negative	In the *positive* direction when the sign is negative

As you can see, vertical transformations are easier to understand. When written in function notation, a vertical transformation occurs *outside* the parentheses.

This placement means that you're just tacking a constant value k onto the end of a function. When k is positive, it moves the graph up k units; when it's negative, it moves the graph down k units.

In contrast, horizontal transformations are a little trickier. In function notation, a horizontal transformation is written *inside* the parentheses.

This placement means that you're adding or subtracting a value h *before* you evaluate the function, which makes the calculation a little more complicated. A positive "h" means the equation simplifies to include $(x-h)$ and is shifted to the right, and a negative "h" means the equation simplifies to include $(x+h)$ and is shifted to the left.

Keeping clear about the differences between vertical and horizontal transformations can often earn you a few extra points on the SAT.

Combining vertical and horizontal transformations

You can combine a vertical and a horizontal transformation to move the graph of a parent function anywhere on the xy-plane. For example:

EXAMPLE

Which of the following transformations moves the graph *left 2 units* and *down 1 unit*?

(A) $f(x+1)+2$

(B) $f(x+2)-1$

(C) $f(x-1)-2$

(D) $f(x-2)+1$

The transformation $f(x+2)$ moves the graph left 2 units. The transformation $f(x)-1$ moves it down 1 unit. Combining these two transformations results in $f(x+2)-1$, so Answer B is correct.

Here's another example that requires you to apply both a vertical and a horizontal transformation to a function:

Which of the following transformations of the function $f(x)=x^2$ results in a function $g(x)$ in which the graph moves *right 1 unit* and *down 4 units*?

(A) $g(x)=x^2+2x+3$

(B) $g(x)=x^2-2x-3$

(C) $g(x)=x^2+2x+5$

(D) $g(x)=x^2-2x-5$

The transformation $f(x-1)-4$ moves the graph right 1 unit and down 4 units. Applying this transformation to $f(x)=x^2$ gives you:

$$f(x-1)-4=(x-1)^2-4$$

To simplify this function, begin by expanding the expression $(x-1)^2$:

$$=(x-1)(x-1)-4$$

Now, FOIL the result and combine like terms:

$$=x^2-x-x+1-4=x^2-2x-3$$

Therefore, Answer B is correct.

Understanding stretch-compress-flip transformations

Multiplying a function by a constant a, thus changing it to $af(x)$, changes its basic shape in a variety of predictable ways, depending on the value of a.

In this section, I show you how multiplying by different values of a affects a function.

When a is positive

When a is *positive*, the transformation $af(x)$:

>> **Stretches the function when a is greater than 1.** For example, $2f(x)$ stretches the function, so that its *y*-values *increase 2 times as quickly* on the *xy*-plane.

>> **Has no effect when a equals 1.** For example, $1f(x) = f(x)$, which leaves the function unchanged.

>> **Compresses the function when a is less than 1.** For example, $\frac{1}{3}f(x)$ compresses the function, so that it *increases one-third as quickly* on the *xy*-plane.

In Figure 10-5, I show you how stretch and compress transformations affect the function $f(x) = x^2$. Here, the graph in the middle is unchanged, because $a = 1$.

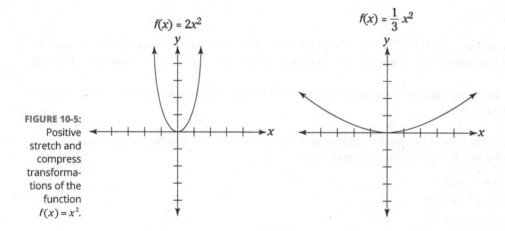

$$f(x) = 2x^2 \qquad f(x) = \frac{1}{3}x^2$$

FIGURE 10-5: Positive stretch and compress transformations of the function $f(x) = x^2$.

You can also apply these transformations to the other parent functions that I discuss earlier in this chapter. For example:

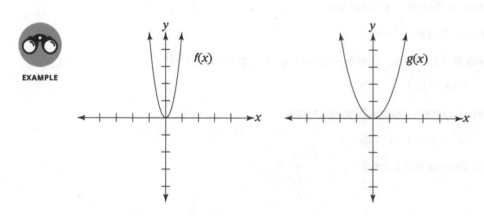

EXAMPLE

$f(x)$ $\qquad$ $g(x)$

The first graph above shows a function $f(x)$. Which of the following could be the equation for the function $g(x)$, as shown in the second graph?

(A) $g(x) = 3f(x)$

(B) $g(x) = -3f(x)$

(C) $g(x) = \frac{1}{3}f(x)$

(D) $g(x) = -\frac{1}{3}f(x)$

The graph of $g(x)$ is a compressed but non-reflected version of $f(x) = f(x)$, so $g(x) = af(x)$ where a is a positive value less than 1. This value could be $\frac{1}{3}$, so Answer C is correct.

When a is negative

When a is negative, the transformation $af(x)$ reflects the function across the x-axis. Additionally, when a is *negative*, the transformation $af(x)$:

> » **Flips *and* stretches the function when a is less than –1.** For example, $-2f(x)$ reflects the function across the x-axis and stretches the function, so that it *decreases 2 times as quickly* on the *xy*-plane.

> » **Only flips the function when a equals –1.** For example, $-f(x)$ reflects the function across the x-axis without stretching or compressing it.

> » **Flips *and* compresses the function when $-1 < a < 0$.** For example, $-\frac{1}{3}f(x)$ reflects the function across the x-axis and compresses the function, so that it *decreases one-third as quickly* on the *xy*-plane.

In Figure 10-6, you can see how three different negative values of a change the graph of the function $f(x) = x^2$. Note that when $a = -1$, the resulting transformation flips the graph but doesn't stretch or compress it.

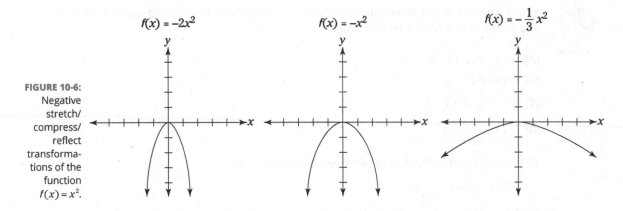

FIGURE 10-6: Negative stretch/ compress/ reflect transformations of the function $f(x) = x^2$.

$f(x) = -2x^2$ $f(x) = -x^2$ $f(x) = -\frac{1}{3}x^2$

You can also apply these types of transformations to other parent functions, such as those that I discuss earlier in this chapter. For example:

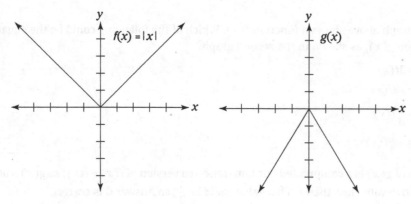

EXAMPLE

In this graph, $f(x) = |x|$. Which of the following could be the equation for the function $g(x)$?

(A) $g(x) = 2|x|$

(B) $g(x) = -2|x|$

(C) $g(x) = \frac{1}{2}|x|$

(D) $g(x) = -\frac{1}{2}|x|$

The graph of $g(x)$ is a reflected and stretched version of $f(x) = |x|$, so $g(x) = a|x|$ where a is a negative value less than 1. This value could be -2, so Answer B is correct.

Working with all three types of transformations

When you're familiar with each of the three most common types of transformations of functions, you're ready to answer more difficult questions that include a variety of transformations in a single function. For example:

EXAMPLE

Which transformation of the function $f(x) = \sqrt{x}$ moves the function left 2 units, down 4 units, and stretches it vertically by a factor of 3?

(A) $f(x) = 3\sqrt{x-4} - 2$

(B) $f(x) = 3\sqrt{x+2} - 4$

(C) $f(x) = \frac{1}{3}\sqrt{x+2} + 4$

(D) $f(x) = \frac{1}{3}\sqrt{x+4} - 2$

The transformation that has the desired effect on $f(x)$ is:

$3f(x+2) - 4$

Applying this transformation to $f(x) = \sqrt{x}$ results in the following function:

$f(x) = 3\sqrt{x+2} - 4$

Therefore, Answer B is correct.

In some cases — especially when working with a quadratic function — a question may require you to simplify your answer. For example:

EXAMPLE

Which transformation of the function $f(x) = x^2$ flips the function across the x-axis, moves it left 3 units, and up 1 unit?

(A) $-x^2 - 6x + 8$

(B) $-x^2 - 6x + 10$

(C) $-x^2 - 6x - 8$

(D) $-x^2 - 6x - 10$

The transformation that has the desired effect on $f(x)$ is:

$$-f(x+3)+1$$

Applying this transformation to $f(x) = x^2$ results in the following function:

$$= -(x+3)^2 + 1$$

Simplify this function by expanding and FOILing $(x+3)^2$ and then combining like terms:

$$= -(x+3)(x+3)+1$$
$$= -(x^2 + 3x + 3x + 9) + 1$$
$$= -(x^2 + 6x + 9) + 1$$
$$= -x^2 - 6x - 9 + 1$$
$$= -x^2 - 6x - 8$$

Therefore, Answer C is correct. In Chapter 12, you work with quadratic functions in greater depth.

Chapter **11**

Polynomials

Polynomials are the most common type of function used in math. Both linear functions ($y = mx + b$) and quadratic functions ($y = ax^2 + bx + c$) are part of the larger set of polynomial functions.

In this chapter, I give you a wide perspective on polynomials. To begin, I show you how to identify a polynomial by its degree, which is the value of its greatest exponent. Next, you discover how to find the *y*-intercept of a polynomial by looking at its constant term. Then, you learn how to figure out the end behavior of polynomials from information found in their leading terms. From there, the focus is on finding the *x*-intercepts of polynomials based on their factored form.

You use all this information to sketch the graphs of polynomials when you're not allowed to use your calculator. From there, you discover how to answer SAT questions that ask you to identify equivalent polynomials. And to finish up, I show you how to use synthetic division to divide one polynomial by another.

Knowing Polynomial Basics

The relatively unhelpful way that polynomials are usually defined is as functions of this form:

$$f(x) = a_n x^n + a_{n-1} x^{n-1} + a_{n-2} x^{n-2} \ldots + a_2 x^2 + a_1 x + a_0$$

A better way to understand them is by relating them to the functions that you're already most familiar with, such as linear and quadratic functions.

In Table 11-1, I list some of the simplest and most common polynomials.

TABLE 11-1   The Polynomials of Degrees 1 through 4

Degree	Polynomial	Parent Function	Standard Form	Example
1	Linear	$y = x$	$y = mx + b$	$y = 2x + 5$
2	Quadratic	$y = x^2$	$y = ax^2 + bx + c$	$y = 4x^2 + 7x - 3$
3	Cubic	$y = x^3$	$y = ax^3 + bx^2 + cx + d$	$y = -x^3 + 5x^2 + x - 5$
4	Quartic	$y = x^4$	$y = ax^4 + bx^3 + cx^2 + dx + e$	$y = 6x^4 + 7x^3 + 8x^2 + 9x - 10$

As you can see in the last column, polynomials are most commonly written in *standard form* — that is, with each term listed in descending order of its exponent. In standard form, the first term of a polynomial — called its *leading term* — tells you its degree, and the last term is the constant, which tells you its y-intercept.

In this section, you work with these two key elements of a polynomial.

Looking at the leading term

The leading term of a polynomial tells you the general shape of its graph when plotted on the xy-plane. It also tells you the positive end behavior of the graph — that is, whether the graph tends toward ∞ or $-\infty$ as x increases.

In this section, I show you how to use the leading term to answer SAT questions.

Identifying the degree of a polynomial

Every polynomial has a *degree*, which is the value of its greatest exponent. For example, the quartic polynomial is degree 4 because its greatest exponent equals 4. In standard form, the degree of a polynomial equals the exponent of its leading term.

The degree of a polynomial gives its general shape when graphed on the xy-plane.

In Chapter 10, I give you a first look at the graphs of the parent functions for polynomials of degrees 1 through 4. In Figure 11-1, I reproduce these graphs for convenience.

As you can see, the degree of a polynomial governs the shape of its graph. Thus, the first step when sketching the graph of a polynomial, as I show you later in this chapter, is to identify its degree.

You can use your understanding of degree to answer some SAT math questions. For example:

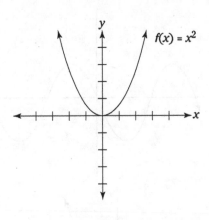

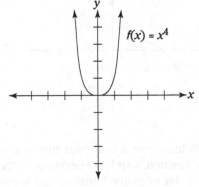

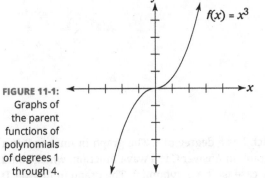

FIGURE 11-1:
Graphs of
the parent
functions of
polynomials
of degrees 1
through 4.

 Which of the following could be the graph of a polynomial whose degree equals 3?

EXAMPLE   **(A)**

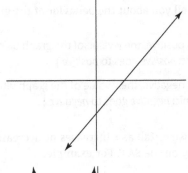

(B)

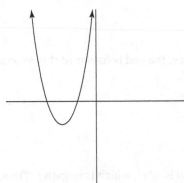

(C)

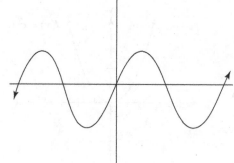

(D)

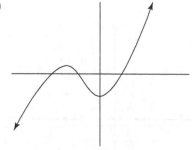

The graph in Answer A is a linear function, which has a degree of 1. The graph in Answer B is a quadratic function, so it has a degree of 2. The graph in Answer C is a wave function, which could be either a sine or cosine function, and in either case isn't a polynomial. The graph in Answer D is a cubic function, whose degree equals 3, so Answer D is correct.

Positive and negative polynomials

The leading term of a polynomial can also tell you about the behavior of the graph as *x* increases:

» When the leading term of a polynomial is positive, the *y*-value of the graph ultimately rises toward ∞ as *x* increases. (Quick mnemonic: *positive* goes to *positive*.)

» When the leading term of a polynomial is negative, the *y*-value of the graph ultimately drops toward −∞ as *x* increases. (Quick mnemonic: *negative* goes to *negative*.)

This tendency for the graph to eventually rise or fall as *x* increases or decreases is called the *end behavior* of the graph, and it comes in handy on the SAT. For example:

$$f(x) = -2x^5 + 3x^4 - x + 1$$

EXAMPLE

When graphed on the *xy*-plane, as *x* increases, the end behavior of this function:

(A) Approaches the value 1.

(B) Approaches the value −2.

(C) Approaches positive infinity (∞).

(D) Approaches negative infinity (−∞).

The leading term of this polynomial function is $-2x^5$, which is negative. Thus, as *x* increases, the end behavior of its graph approaches negative infinity (−∞), so Answer D is correct.

Using the constant term to find the *y*-intercept

The last term of a polynomial written in standard form is its *constant term* — that is, the term that doesn't include a variable, such as *x*. This term is always the *y*-intercept of any polynomial — that is, the point where the graph crosses the *y*-axis.

WARNING

When a polynomial doesn't appear to have a constant term, its constant term equals 0.

You're probably already familiar with this feature when working with linear functions: In a function of the form $y = mx + b$, the constant b is the *y*-intercept. (Flip to Chapter 5 if you're not clear about this idea.)

This connection between the constant and the *y*-intercept works for any polynomial function, and it can help you answer SAT questions. For example:

EXAMPLE

$$f(x) = x^4 + 2x^3 + 5$$

When graphed on the *xy*-plane, what is the *y*-intercept of this function?

The function is a polynomial, so its *y*-intercept equals the value of its constant term. Therefore, the answer is 5.

Identifying the *y*-intercept can be a quick way to gain a point on the SAT. You'll also use it later in this chapter when sketching the graph of a polynomial.

Identifying Odd and Even Polynomials

In the previous section, you discovered how to use the leading term of a polynomial to identify its basic shape and its end behavior as *x* increases toward ∞. In this section, you discover how to tell whether a polynomial is odd or even. Then you use this information to find out its end behavior as *x* decreases toward –∞.

Identifying odd and even functions

Recall that the leading term of a polynomial is the term with the greatest exponent. You can use this term to identify whether any polynomial is an *odd polynomial* or an *even polynomial*:

» If the exponent of the leading term is an odd number, the polynomial is odd.

» If the exponent of the leading term is an even number, the polynomial is even.

You can use this information to answer SAT math questions. For example:

EXAMPLE

$$f(x) = -x^{23} + 2x^{12} + 3x^2 - 9x$$

Which of the following is a true statement about this polynomial?

(A) This is an odd polynomial because its leading term is negative.

(B) This is an even polynomial because its leading term is negative.

(C) This is an odd polynomial because its leading term has an odd exponent.

(D) This is an even polynomial because it has an even number of terms.

The exponent of this polynomial is 23, which is an odd number, so the polynomial is odd. Therefore, Answer C is correct.

Spotting end behavior as *x* approaches −∞

Earlier in this chapter, you discovered how to determine the end behavior of any polynomial as *x* increases, including mnemonics to help you remember. You can use this information as a stepping-stone for finding the end behavior of a polynomial when *x* decreases.

First, here's some quick information to remember:

» When a polynomial is odd, its end behavior goes in *opposite* directions. (Quick mnemonic: *O* is the first letter of both *odd* and *opposite*.)

» When a polynomial is even, its end behavior goes in the *same* direction. (Quick mnemonic: The words *even* and *same* both have four letters.)

Now, here's how to use this information when studying a specific polynomial.

1. **Check whether the leading coefficient of the polynomial is positive or negative:**

   - If it's positive, then as *x* increases, the graph goes *upward*.
   - If it's negative, then as *x* increases, the graph goes *downward*.

2. **Check whether the degree of the polynomial is odd or even:**

   - If it's *odd*, then as *x* decreases, the graph goes in the *opposite* direction.
   - If it's *even*, then as *x* decreases, the graph goes in the *same* direction.

Table 11-2 encapsulates this information for all four possible cases.

TABLE 11-2 **End Behavior of Polynomial Functions as $x \to \infty$ and as $x \to -\infty$**

Polynomial Type	Positive	Negative
Odd (opposite directions)	As $x \to \infty, f(x) \to \infty$	As $x \to \infty, f(x) \to -\infty$
	As $x \to -\infty, f(x) \to -\infty$	As $x \to -\infty, f(x) \to \infty$
Even (same direction)	As $x \to \infty, f(x) \to \infty$	As $x \to \infty, f(x) \to -\infty$
	As $x \to -\infty, f(x) \to \infty$	As $x \to -\infty, f(x) \to -\infty$

I'm discussing this concept from a number of possible angles because you may find it a bit confusing. Here are a couple of examples to drive the point home:

EXAMPLE

$$y = x^3 + 6x^2 + 4x - 9$$

Which of the following accurately describes the end behavior of this polynomial function?

(A) As x increases, y approaches ∞; as x decreases, y approaches ∞.

(B) As x increases, y approaches ∞; as x decreases, y approaches $-\infty$.

(C) As x increases, y approaches $-\infty$; as x decreases, y approaches ∞.

(D) As x increases, y approaches $-\infty$; as x decreases, y approaches $-\infty$.

The function is positive, so as x increases, y approaches ∞. And the function is odd, so as x decreases, y goes in the opposite direction, so it approaches $-\infty$. Therefore, Answer B is correct.

EXAMPLE

Which of the following is an example of a function that ultimately approaches $-\infty$ both as x increases and as x decreases?

(A) $f(x) = 4x^3 - 2$

(B) $f(x) = -x^4 - 8x^3 + 6x^2 - 9x$

(C) $f(x) = -2x^5 + x + 1$

(D) $f(x) = 5x^6 - x^5 - 10x^4 + 3x^3 + x - 5$

A function that approaches $-\infty$ as x increases has a negative leading coefficient, so you can rule out Answers A and D. If this function also approaches $-\infty$ as x decreases, its end behavior is the same in both directions; therefore, it's even, which rules out Answer A. Thus, Answer B is correct.

I hope these examples help you to see that this information is a lot easier to use when you have a specific problem to solve. And that's where you'll be when you take your SAT.

Finding the *x*-Intercepts of Polynomials

The x-intercepts of a polynomial are the points where its graph crosses the x-axis. Unlike the y-intercept, which is easy to find for every polynomial (just check the constant term!), x-intercepts can be tricky, because they may be numerous or even non-existent.

In this section, I show you how to distinguish x-intercepts from roots, solutions, and zeros, all words that the SAT uses relatively interchangeably when discussing polynomials. You also discover how to find the x-intercepts of a polynomial in factored form. Finally, you see how to find the x-intercepts of a polynomial in standard form by factoring it first.

Knowing a few names for *x*-intercepts

The x-intercepts of a graph are points (if any) where that graph intersects the x-axis. On the SAT, you may contend with a variety of names for x-intercepts, such as:

>> Real roots

>> Real solutions

>> Real zeros

Why do mathematicians need all these different terms to describe a single idea? Job security!

I like the term *x-intercepts* because it's very clear, and most of my students immediately get what I'm talking about.

Here's another reason why I like it: The words *roots*, *solutions*, and *zeros* refer to essentially the same idea as *x*-intercepts. However, there's a catch: Depending on the context, all three of those words can also include *complex values* — that is, values that aren't on the number line because they contain an imaginary part $i = \sqrt{-1}$.

That's why in the list of names, I purposely write *real roots*, *real solutions*, and *real zeros*. And that's why I prefer to call them by the name *x*-intercepts, which refers only to *real* values of *x* — that is, values where the graph of the polynomial *really* crosses the *x*-axis.

I touch briefly upon complex roots of a polynomial in Chapter 12, where I discuss quadratic functions.

Finding the *x*-intercepts of a polynomial in factored form

Finding the *x*-intercepts of a polynomial *with* a calculator is a good skill to have for the calculator portion of the SAT. In Chapter 2, I show you how to do this.

To find the *x*-intercepts of a polynomial *without* a calculator, factor it, then set each resulting factor to 0 and solve for *x*. This process is simplest when the polynomial is factored to begin with. For example:

EXAMPLE

$$f(x) = 3x(x+2)(x^2+4)(2x-5)$$

If $k > 0$ is a real root of this polynomial, what is the value of *k*?

The polynomial is given in factored form, so set each factor equal to 0 and solve for *x*:

$$3x = 0 \qquad x + 2 = 0 \qquad x^2 + 4 = 0 \qquad 2x - 5 = 0$$
$$x = 0 \qquad x = -2 \qquad x^2 = -4 \qquad 2x = 5$$
$$x = \pm\sqrt{-4} = \pm 2i \qquad x = \frac{5}{2}$$

The first two factors result in the *x*-intercepts 0 and −2, both of which are incorrect because $k \geq 0$. The third factor results in the pair of complex roots $\pm 2i$, so this is also incorrect. The fourth factor results in the *x*-intercept $\frac{5}{2}$, so the correct answer is either $\frac{5}{2}$ or 2.5.

Finding the *x*-intercepts of a polynomial in standard form

When a polynomial is in standard form — that is, not factored — finding the roots without a calculator usually depends on your finding a way to factor it. In Chapter 3, I show you a variety of ways to factor polynomials. In the next example, I factor a cubic polynomial using the method shown there:

EXAMPLE

$$f(x) = 10x^3 - 3x^2 + 20x - 6$$

If this function intersects $(n,0)$, what is the value of n?

The point $(n,0)$ is an x-intercept. Begin finding this point by factoring the cubic expression by grouping:

$$\begin{aligned} f(x) &= 10x^3 - 3x^2 + 20x - 6 \\ &= x^2(10x - 3) + 2(10x - 3) \\ &= (x^2 + 2)(10x - 3) \end{aligned}$$

Now, set each factor equal to 0:

$$\begin{array}{ll} x^2 + 2 = 0 & 10x - 3 = 0 \\ x^2 = -2 & 10x = 3 \\ x = \pm\sqrt{-2} = \pm\sqrt{2}i & x = \dfrac{3}{10} \end{array}$$

The first factor results in the pair of complex roots $\pm i$. The second factor results in the value $\dfrac{3}{10}$, so the correct answer is either $\dfrac{3}{10}$ or .3.

Sketching the Graph of a Polynomial

When you don't have access to a graphing calculator — as is the case when you take the No Calculator section of the SAT — producing a quick sketch of a polynomial graph can be a life saver.

Throughout this chapter, I've presented a bunch of tools for understanding polynomial functions. In this section, you put all these tools to work as you discover how to sketch the graphs of a polynomial.

Your sketch doesn't have to be perfect, but to be useful, it needs to accurately reflect as many of the following features as you can find:

>> The y-intercept

>> End behavior as x increases and decreases

>> All x-intercepts (if any!)

>> Cross-through versus bounce x-intercepts (which I explain next)

Having the skills to draw a quick but accurate sketch of a polynomial increases your chances of answering even difficult questions that you may encounter.

Sketching a polynomial from its factored form

When you're sketching the graph of a polynomial presented in factored form, the x-intercepts should be relatively easy to find, so graph these first. After that, to complete your sketch, you'll want to find the y-intercept and the end behavior. A quick way to do this is to:

>> Multiply all the x-terms to find the leading term of the polynomial.

>> Multiply all the constant terms to find the constant term of the polynomial.

Here's an example to help make sense of this process:

$$f(x) = (x-1)(x-3)(x+4)$$

EXAMPLE   Which of the following is the most accurate sketch of this polynomial?

(A)

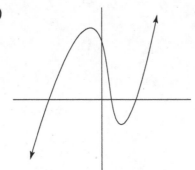

(B)

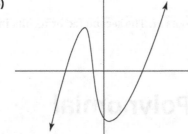

(C)

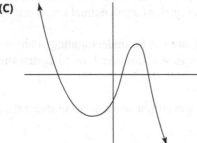

(D)

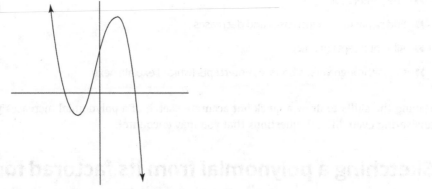

This polynomial has x-intercepts at 1, 3, and −4, so Answers B and D are ruled out. To find the y-intercept and the end behavior, find the leading and constant terms by multiplying the three x-terms and the three constant terms:

$$x \cdot x \cdot x = x^3 \qquad\qquad -1 \cdot (-3) \cdot 4 = 12$$

These two values give you the leading and constant terms of the polynomial:

$$f(x) = (x-1)(x-3)(x+4) = x^3 \ldots + 12$$

Even though you haven't found the middle terms of the polynomial, you have enough information to answer the question. This information shows that the polynomial has a y-intercept of 12 and end behavior that's positive as x approaches ∞ and negative as x approaches $-\infty$, so Answer A is correct.

Sketching "bounce" x-intercepts

So far in this section, you've worked with x-intercepts that have been unique within a single polynomial — that is, no factor as a whole has been raised to an exponent.

In some cases, however, an SAT question may present you with a polynomial that has a factor raised to an even exponent. This type of exponent results in an x-intercept whose graph "bounces" off the x-axis rather than passing through.

To see how this difference looks on the xy-graph, look at the graph of the polynomial $f(x) = (x+1)^2(x-2)$ shown in Figure 11-2. When completely factored, this polynomial is $f(x) = (x+1)(x+1)(x-2)$, with the factor $(x+1)$ appearing twice. Thus, this graph has x-intercepts at both 2 and -1. As you can see, the graph crosses the x-intercept at 2 because the corresponding factor, $(x-2)$, isn't raised to a power. However, the graph bounces off the x-intercept at -1, touching this point without passing through. This behavior happens because the corresponding factor, $(x+1)$, is raised to the power of 2, which is an even positive number.

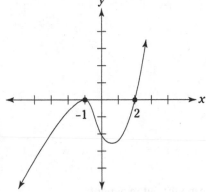

FIGURE 11-2:
Comparing cross-through and bounce intercepts.

Here's an example of an SAT math question that draws from this information:

EXAMPLE

$$f(x) = (x+1)(x-2)^2$$

Which of the following could be a reasonably accurate sketch of this polynomial function?

(A)

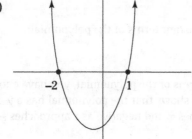

(B)

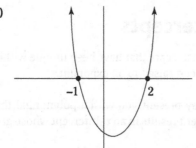

(C)

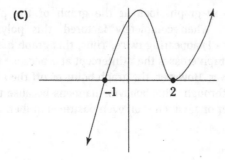

(D)

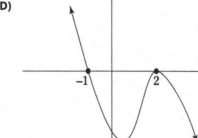

To begin, expand this polynomial to factored form without exponents:

$$f(x) = (x+1)(x-2)(x-2)$$

Now, set each distinct factor equal to 0:

$x+1=0$ $x-2=0$

$x=-1$ $x=2$

Thus, this polynomial has two distinct x-intercepts at −1 and 2. At −1, the graph passes through the x-axis. However, at 2, the graph bounces off the x-axis without passing through, so Answers A and B are both ruled out.

To narrow down the answer further, find the leading term and constant term of the function:

$$f(x) = (x+1)(x-2)(x-2) = x^3 + \ldots + 4$$

Thus, this polynomial has a y-intercept of 4, which rules out Answer D, so Answer C is correct.

Sketching a polynomial from its standard form

When you're sketching the graph of a polynomial that's in standard form, start with the y-intercept and end behavior, because these should be relatively easy to find. In some cases, this information will be enough to answer a question. For example:

$$f(x) = x^4 + kx^2 - n$$

EXAMPLE

If k and n are both positive integers, which of the following could be a reasonably accurate sketch of this polynomial?

(A)

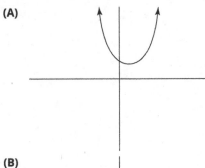

(B)

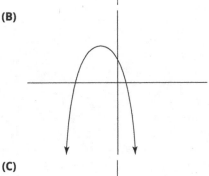

(C)

(D)

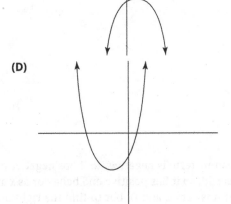

The function has a y-intercept of $-n$, which is below the x-axis, so this rules out Answers A and B. Its leading term is positive and even, so it has positive end behavior in both directions, which rules out Answer C. Therefore, Answer D is correct.

In some cases, you may need more information about the graph than standard form can provide. For example:

$$f(x) = -2x^3 + 6x^2 + 20x$$

EXAMPLE Which of the following is the most accurate graph of this polynomial function?

(A)

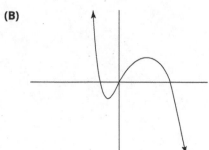

(B)

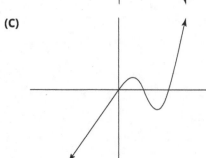

(C)

(D)

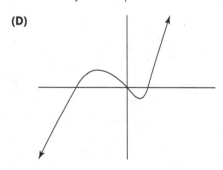

This polynomial has a y-intercept of 0. Its leading term is negative, so it has negative end behavior as x approaches ∞. And its leading term is odd, so it has positive end behavior as x approaches $-\infty$. This information is sufficient to rule out Answers C and D, but to find the right answer, you need to find the x-intercepts.

Fortunately, this polynomial is factorable, so begin by factoring out the GCF:

$$f(x) = -2x(x^2 - 3x - 10)$$

The result includes a quadratic factor that is also factorable:

$$= -2x(x + 2)(x - 5)$$

The resulting polynomial has roots at −2, 0, and 5. This result rules out Answer A, which has two negative x-intercepts, so Answer B is correct.

Equivalent Polynomials

A variety of SAT questions — ranging from easy to medium to hard — hinge on the following theorem about polynomials:

REMEMBER

When a pair of polynomials are equivalent for all values of x, their corresponding coefficients are also equivalent.

To get a sense of what this theorem is talking about, consider the following SAT question:

EXAMPLE

If $(x + 3)(x + 6) = ax^2 + bx + c$ is true for all values of x, what is the value of $a + b + c$?

To solve this problem, begin by FOILing the left side of the equation:

$$x^2 + 9x + 18 = ax^2 + bx + c$$

Because they're equal, the two polynomials in this equation must have corresponding coefficients that are also equivalent, so $a = 1$, $b = 9$, and $c = 18$. Thus, $a + b + c = 28$.

This same basic idea can give rise to problems that are a bit more confusing. For example:

EXAMPLE

If $x(x - 1) + 3(x - 1) + 13 = x^2 + kx + kn$ for all values of x, what is the value of $k + n$?

To begin, simplify the left side of the equation by distributing and combining like terms:

$$x^2 - x + 3x - 3 + 13 = x^2 + kx + kn$$
$$x^2 + 2x + 10 = x^2 + kx + kn$$

With both polynomials in standard form, you can set the coefficients for the two x-terms (2 and k) equal to each other, so:

$$k = 2$$

Additionally, you can set the two constants equal to each other, so:

$$kn = 10$$

To solve for n, plug in 2 for k:

$$2n = 10$$
$$n = 5$$

Thus, $k + n = 2 + 5 = 7$, so the correct answer is 7.

Recalling Synthetic Division for Polynomials

Most SAT students have a vague memory of having done polynomial division at some time in the past. For whatever reason, it just doesn't seem to stick.

Fortunately, for the purposes of the SAT, you can probably get away with a simple refresher on synthetic division, which is sort of polynomial division *lite*. An example should help you remember:

EXAMPLE

Which of the following expressions is equivalent to $\dfrac{x^2 - 3x + 8}{x + 2}$?

(A) $x - 5 + \dfrac{2}{x + 2}$

(B) $x - 5 + \dfrac{7}{x + 2}$

(C) $x - 5 + \dfrac{9}{x + 2}$

(D) $x - 5 + \dfrac{18}{x + 2}$

To solve this problem, divide $x^2 - 3x + 8$ by $x + 2$ using synthetic division. To begin, pull out the three coefficients of the polynomial $x^2 - 3x + 8$ (1, −3, and 8), and flip the sign of the constant $x + 2$ (that is, change 2 to −2). Arrange these numbers as follows

$$-2 \,\big|\, \begin{array}{ccc} 1 & -3 & 8 \end{array}$$

To begin, bring down the number 1 as follows:

$$-2 \,\big|\, \begin{array}{ccc} 1 & -3 & 8 \\ \hline 1 \end{array}$$

Next, multiply $1 \times (-2) = -2$, and place this value above the line in the next column:

$$-2 \,\big|\, \begin{array}{ccc} 1 & -3 & 8 \\ & -2 & \\ \hline 1 \end{array}$$

Now, add $-3 + (-2) = -5$, and place this sum at the bottom of the next column:

$$-2 \,\big|\, \begin{array}{ccc} 1 & -3 & 8 \\ & -2 & \\ \hline 1 & -5 \end{array}$$

Now, repeat these three steps, multiplying $-5 \times (-2) = 10$, placing this value below the 8, and then adding $8 + 10 = 18$:

$$-2 \,\big|\, \begin{array}{ccc} 1 & -3 & 8 \\ & -2 & 10 \\ \hline 1 & -5 & |18 \end{array}$$

To interpret the result, consider the values 1 and −5 as the coefficients of the resulting polynomial $x - 5$, and the value 18 as the remainder. Place this remainder in the numerator of a rational expression with the original denominator $x + 2$ and attach it to the polynomial you found, as follows:

$$x - 5 + \dfrac{18}{x + 2}$$

Thus, the original expression $\dfrac{x^2 - 3x + 8}{x + 2}$ equals $x - 5 + \dfrac{18}{x + 2}$. Therefore, Answer D is correct.

IN THIS CHAPTER

» Culling information from a quadratic function in standard form: $f(x) = ax^2 + bx + c$

» Understanding how to set up a quadratic function in vertex form: $f(x) = a(x - h)^2 + k$

» Converting a function from standard to vertex form, and vice versa

» Finding the roots of a quadratic function

» Solving word problems about projectiles with quadratic functions

Chapter **12**

Quadratic Functions

Q uadratic functions are a staple of any algebra class, and among the most common topics to be found on the SAT.

In this chapter, I show you how to understand two important forms of the quadratic function: standard form, $f(x) = ax^2 + bx + c$, and vertex form, $f(x) = a(x - h)^2 + k$. Next, you discover how to change an equation in vertex form to standard form, and vice versa.

After that, you find out how to find the roots (also called *solutions*, *x-intercepts*, and *zeros*) of a quadratic function by setting it equal to 0 and solving the resulting quadratic equation for *x*. And to finish up, you solve projectile problems, which are a common type of SAT word problem that relies on a knowledge of quadratic functions.

By the way, throughout this chapter, I use the variables *y* and *f(x)* interchangeably, just as they are on the SAT. Don't let the notation confuse you! The functions $y = ax^2 + bx + c$ and $y = a(x - h)^2 + k$ are simply alternate forms of those that use functional notation *f(x)*.

The Quadratic Function in Standard Form

You're probably most familiar with quadratic functions in standard form:

$$f(x) = ax^2 + bx + c$$

In this section, I show you a variety of ways to squeeze as much helpful information as you need out of a quadratic function in standard form, focusing on the values a, b, and c. After that, I show you how to use this information to sketch quadratics.

Understanding standard-form quadratic functions

You can find out a lot of information about a quadratic function by examining it in standard form. In this section, you discover four key types of information that are most helpful for answering SAT questions:

>> Concavity (concave up or concave down)

>> The y-intercept

>> The axis of symmetry

>> The coordinates of the vertex

This information should enable you to sketch a useful graph of just about any quadratic function.

Identifying concavity

The leading term of a quadratic function in standard form, $f(x) = ax^2 + bx + c$, includes the coefficient a. This coefficient tells you whether the function is *concave up* or *concave down* (also called *convex*), as shown in Figure 12-1:

>> When a is positive, the function is concave up.

>> When a is negative, the function is concave down.

TIP

A mnemonic to remember this rule is that a concave up quadratic function looks like a smiley face, and a concave down quadratic function looks like a frowny face. Simple, right?

This rule makes sense when you think about it. Quadratic functions are *even* polynomials. As you know from Chapter 11, when an even polynomial has a positive leading coefficient, its end behavior in both directions tends toward positive infinity. Similarly, when it has a negative coefficient, its end behavior tends toward negative infinity.

Also, keep in mind the following:

>> When a quadratic function is concave up ($a > 0$), it has a *minimum* value at its vertex.

>> When a quadratic function is concave down ($a < 0$), it has a *maximum* value at its vertex.

This information about concave-up and concave-down quadratic functions is encapsulated in Figure 12-1.

FIGURE 12-1:
Positive
quadratic
functions
are concave
up and have
a minimum
value.
Negative
quadratic
functions
are concave
down
(convex)
and have a
maximum
value.

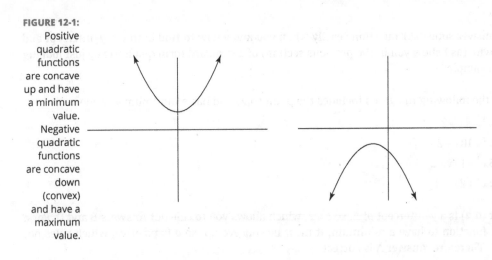

Information about the concavity of a quadratic function can be useful when answering SAT math questions. For example:

EXAMPLE

If $f(x) = 3x^2 + 5x - 4$ and $g(x) = -2x^2 + 6x - 1$, then which of the following is true?

(A) $f(x)$ and $g(x)$ both have maximum values.

(B) $f(x)$ has a maximum value and $g(x)$ has a minimum value.

(C) $f(x)$ has a minimum value and $g(x)$ has a maximum value.

(D) $f(x)$ and $g(x)$ both have minimum values.

The leading coefficient of $f(x)$ is 3, which is positive, so $f(x)$ is concave up and has a minimum value. In contrast, the leading coefficient of $g(x)$ is –2, which is negative, so $g(x)$ is concave down and has a maximum value. Therefore, Answer C is correct.

Plotting the *y*-intercept

The last term of a quadratic function in standard form is the constant c, so this term tells you the y-intercept of the function, which is $(0, c)$. This intersection occurs because $x = 0$ where the function crosses the y-axis. Thus, the first two terms of the function $y = ax^2 + bx + c$ also equal 0, so $y = c$. You see a couple of examples in Figure 12-2.

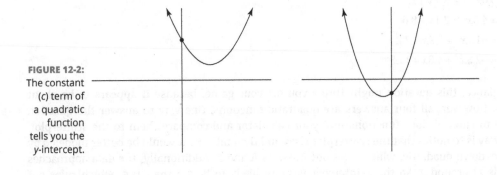

FIGURE 12-2:
The constant
(*c*) term of
a quadratic
function
tells you the
y-intercept.

You can answer some SAT questions easily when you know how to find both the y-intercept and the concavity (as I show you in the previous section) of a standard form quadratic by just looking at it. For example:

EXAMPLE

Which of the following functions includes the point $(0,2)$ and has a minimum y-value?

(A) $y = 3x^2 - 8x + 2$

(B) $y = 2x^2 - 10x - 2$

(C) $y = -5x^2 - 15x + 2$

(D) $y = -2x^2 + 2x - 3$

The point $(0,2)$ is a y-intercept of 2, so $c = 2$, which allows you to rule out Answers B and D. For a quadratic function to have a minimum, it must be concave up, so a is positive, which rules out Answer C. Therefore, Answer A is correct.

Here's another example of an entirely different type of question that uses basic information from a quadratic function in standard form.

EXAMPLE

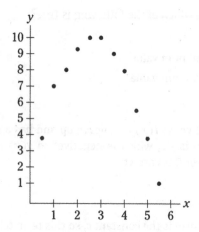

This scatterplot shows statistical data about a deer population over a six-year period. The x-value represents the number of years after 2015. The y-value represents the number of deer in thousands. Which of the following functions best models this data set?

(A) $f(x) = 0.8x^2 + 8.7x + 1.9$

(B) $f(x) = 4.5x^2 + 3.1x + 8.6$

(C) $f(x) = -1.1x^2 + 5.9x + 2.2$

(D) $f(x) = -2.3x^2 + 4.3x + 11.5$

At first glance, this question might throw you off your game, because it appears to be about statistics. However, all four answers are quadratic functions. One way to answer the question might be to enter all four functions into your calculator and compare them to the scatterplot. A faster way is to notice that the scatterplot rises and then falls, so it would be better modeled by a concave-down quadratic, which rules out Answers A and B. Additionally, the data approaches the y-axis at around 2, so the y-intercept is more likely to be 2.2 than 11.5, which rules out Answer D. Thus, Answer C is correct.

Calculating the axis of symmetry

On the xy-plane, the graph of every quadratic function takes the shape of a parabola, which is left-right symmetrical. That is, if you were to place a mirror vertically down the middle of a parabola, the reflection would take the same path as the function itself.

This symmetry allows you to draw a vertical line down the middle of every parabola, called the *axis of symmetry*. The formula for the axis of symmetry is:

$$x = -\frac{b}{2a}$$

Here's a sample question where this formula is useful:

EXAMPLE

The function $f(x) = 3x^2 - 12x + 5$ is symmetrical across the vertical line on the xy-plane where $x =$

(A) -4

(B) -2

(C) 2

(D) 4

In this function, $a = 3$ and $b = -12$. Plug these values into the formula for the axis of symmetry as follows:

$$x = -\frac{-12}{2(3)} = \frac{12}{6} = 2$$

Thus, the vertical line $x = 2$ is the axis of symmetry for this function, so Answer C is correct.

Here's a question that uses information about concavity, the y-intercept, and the axis of symmetry:

EXAMPLE

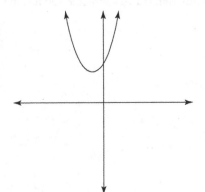

Which of the following could be the function for this graph?

(A) $f(x) = 2x^2 + 3x + 5$

(B) $g(x) = 2x^2 + 3x - 5$

(C) $h(x) = 2x^2 - 3x + 5$

(D) $j(x) = -2x^2 - 3x + 5$

The graph shown is concave up, which rules out Answer D. It also has a positive y-intercept, which rules out Answer B. Its axis of symmetry is to the left of the y-axis, so $-\frac{b}{2a}$ is a negative number, which rules out Answer C. Therefore, Answer A is correct.

Finding the vertex

The *vertex* of a parabola is the point where it intersects with its axis of symmetry. The vertex is the minimum y-value (lowest point) on a positive quadratic function, and the maximum y-value (highest point) on a negative quadratic function.

When you know the equation for the axis of symmetry — which I show you how to calculate in the previous section — you can find the x and y values of the vertex as follows:

>> The x-value of the vertex is the same as the axis of symmetry.

>> The y-value of the vertex is the result when you plug the x value into the quadratic and solve.

For example:

EXAMPLE

If (h,k) are the coordinates of the minimum point of the function $y = 3x^2 - 12x + 5$, then

(A) $k < -4$

(B) $-4 <= k < 0$

(C) $0 <= k <= 4$

(D) $k > 4$

In the previous section, you calculated the axis of symmetry for this function as $x = 2$, so this is the h-value of the vertex. To find the value of k, plug in 2 for x into the function and solve:

$$k = 3(2)^2 - 12(2) + 5 = 12 - 24 + 5 = -7$$

Thus, the vertex of this function is at $(2, -7)$, and so $k = -7$; thus, Answer A is correct.

Here's another example:

EXAMPLE

When graphed on the xy-plane, the vertex of the function $f(x) = -x^2 - 10x - 20$ lies in which of the following quadrants?

(A) Quadrant 1

(B) Quadrant 2

(C) Quadrant 3

(D) Quadrant 4

Begin by calculating the axis of symmetry:

$$x = -\frac{b}{2a} = -\frac{-10}{2(-1)} = -\frac{-10}{-2} = -5$$

Now, plug this value into the function:

$$-(-5)^2 - 10(-5) - 20 = -25 + 50 - 20 = 5$$

Thus, the vertex of this function is at $(-5, 5)$, which is in Quadrant 2 (see Chapter 2 for a refresher on the four quadrants of the xy-plane, so Answer B is correct.

Sketching a parabola from standard form

In the previous section, you discovered that standard form allows you to find four attributes of a quadratic function:

>> Concavity (concave up or concave down)

>> The y-intercept

>> The axis of symmetry

>> The coordinates of the vertex

Knowing these four attributes allows you to draw a quick and fairly accurate sketch of a quadratic function.

EXAMPLE

For example, consider the function $f(x) = x^2 - 6x + 8$. In this function, $a = 1$, $b = -6$, and $c = 8$, so you can find these four attributes as follows.

» Concavity: The a-value is 1, which is positive, so the function is concave up.

» The y-intercept: The c-value is 8, so this is the y-intercept.

» The axis of symmetry: Plug a and b into the formula for the axis of symmetry.

$$x = -\frac{b}{2a} = -\frac{-6}{2(1)} = 3$$

Thus, the axis of symmetry is $x = 3$.

» The coordinates of the vertex: The x-value of the vertex is 3, so plug in this value for x into the function to find the y-value.

$$y = (3)^2 - 6(3) + 8 = 9 - 18 + 8 = -1$$

Thus, the vertex is at $(3, -1)$.

These four attributes allow you to sketch a relatively accurate graph of the function:

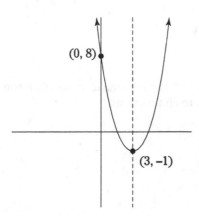

As another example, consider the function $f(x) = -3x^2 + 6x + 2$. In this case, $a = -3$, $b = 6$, and $c = 2$, so you can discover these four attributes as follows.

» Concavity: The a-value is –3, which is negative, so the function is concave down.

» The y-intercept: The c-value is 2, so this is the y-intercept.

» The axis of symmetry: Plug a and b into the formula for the axis of symmetry.

$$x = -\frac{b}{2a} = -\frac{6}{2(-3)} = -\frac{6}{-6} = 1$$

Thus, the axis of symmetry is $x = 1$.

» The coordinates of the vertex: The x-value of the vertex is 1, so plug this value for x into the function to find the y-value.

$$y = -3(1)^2 + 6(1) + 2 = -3 + 6 + 2 = 5$$

Thus, the vertex is at $(1, 5)$.

With this information, you can sketch the following graph:

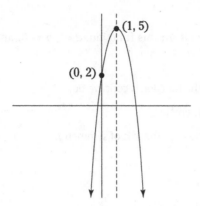

The Quadratic Function in Vertex Form

The vertex form of a quadratic function is $f(x) = a(x-h)^2 + k$. As with the standard form, this form yields a lot of information when you know how to find it. In this section, I first bring you up to speed on how to understand and work with the variables a, h, and k. Then, you discover how to make a rough sketch of a quadratic function in vertex form.

Understanding vertex form quadratic functions

A quadratic function in vertex form $f(x) = a(x-h)^2 + k$ starts to make sense when you think of it in terms of three transformations that are general to all functions.

» Vertical transformations: $f(x) + k$

» Horizontal transformations: $f(x-h)$

» Stretch/compress/reflect transformations: $af(x)$

For a refresher on how these transformations operate on functions in general, turn to Chapter 10. In this section, I focus exclusively on how to apply them to the parent function $f(x) = x^2$.

Vertical and horizontal transformations with *k* and *h*

The vertical and horizontal transformations allow you to move the vertex of the parent function $y = x^2$ to any point on the graph. In this section, you focus on these two transformations, first by themselves and then together.

THE VERTICAL TRANSFORMATION WITH VARIABLE *K*

In Chapter 10, you work with the vertical transformation $f(x) + k$. This transformation does the following:

» **Moves *f(x)* up when the sign is positive:** For example, $f(x) + 1$ moves the function *up 1 unit* on the *xy*-plane.

>> **Moves f(x) down when the sign is negative:** For example, $f(x) - 4$ moves the function *down* 4 *units* on the *xy*-plane.

Figure 12-3 shows these two vertical transformations applied to the quadratic parent function $f(x) = x^2$.

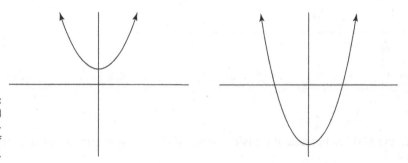

FIGURE 12-3:
Two vertical transforma-
tions of
$f(x) = x^2$.

EXAMPLE

If $f(x) = x^2$ and $g(x)$ is a vertical transformation of $f(x)$ whose minimum value is at $(0, -3)$, then $g(x) =$

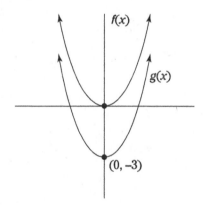

(A) $(x+3)^2$

(B) $(x-3)^2$

(C) $x^2 + 3$

(D) $x^2 - 3$

The vertical transformation from $f(x)$ to $g(x)$ moves the graph down 3 units, so $g(x) = x^2 - 3$; therefore, Answer D is correct.

THE HORIZONTAL TRANSFORMATION WITH VARIABLE *H*

In Chapter 10, I show you the horizontal transformation $f(x - h)$. Note the presence of the minus sign in this transformation, which moves the function opposite from the direction that makes intuitive sense. That is, this transformation:

>> **Moves f(x) to the left when the sign is positive.** For example, $f(x + 2)$ moves the function *left 2 units* on the *xy*-plane — that is, two units in the negative direction.

>> **Moves f(x) to the right when the sign is negative.** For example, $f(x - 3)$ moves the function *right 3 units* on the *xy*-plane — that is, three units in the positive direction.

Figure 12-4 shows these two horizontal transformations applied to the quadratic parent function $f(x) = x^2$.

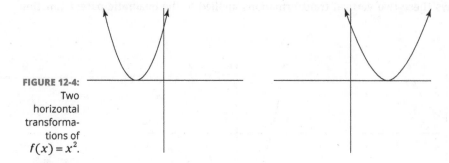

FIGURE 12-4:
Two
horizontal
transforma-
tions of
$f(x) = x^2$.

EXAMPLE

If $f(x) = x^2$ and $h(x)$ is a horizontal transformation of $f(x)$ whose minimum value is at (2,0), then $h(x) =$

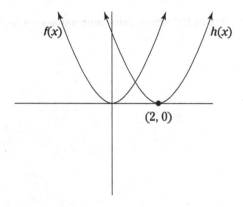

$f(x)$ $h(x)$

(2, 0)

(A) $(x+2)^2$

(B) $(x-2)^2$

(C) $x^2 + 2$

(D) $x^2 - 2$

The vertical transformation from $f(x)$ to $h(x)$ moves the graph right 2 units, so $h(x) = (x-2)^2$; therefore, Answer B is correct.

COMBINING VERTICAL AND HORIZONTAL TRANSFORMATIONS

When you combine a vertical transformation with a horizontal transformation, you can move the parent function $f(x) = x^2$ anywhere on the graph. The easiest point to track is the vertex of the parabola: Applying both a vertical and a horizontal transformation moves the vertex to the point (h,k). For example:

» $f(x+2)+1$ **moves the vertex to (-2,1) on the *xy*-plane.**

» $f(x-3)-4$ **moves the vertex to (3,-4) on the *xy*-plane.**

Figure 12-5 shows these two combinations of vertical and horizontal transformations for $f(x) = x^2$.

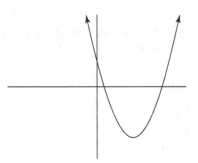

FIGURE 12-5:
Two combi-
nations of
vertical and
horizontal
transforma-
tions of
$f(x) = x^2$.

EXAMPLE

If $f(x) = x^2$ and $j(x)$ is a transformation of $f(x)$ whose vertex is at $(-1, -3)$, then which of the following could be true?

(A) $j(x) = (x+1)^2 + 3$

(B) $j(x) = (x-1)^2 + 3$

(C) $j(x) = (x+1)^2 - 3$

(D) $j(x) = (x-1)^2 - 3$

The transformation from $f(x)$ to $j(x)$ moves the parabola left 1 unit and down 3 units, so $j(x) = (x+1)^2 - 3$; therefore, Answer C is correct.

Stretch-compress-reflect transformations with variable *a*

The variable *a* governs a set of transformations that can stretch or compress the function and, when negative, reflect it across the *x*-axis from concave up to concave down. In this section, you discover how these transformations happen, first one at a time and then when combined.

THE STRETCH-COMPRESS TRANSFORMATION WITH VARIABLE *A*

In Chapter 10, you discover the stretch-compress transformation $af(x)$. When *a* is *positive*, the transformation $af(x)$ does the following:

» **Stretches the function when *a* is greater than 1.** For example, $2f(x)$ stretches the function, so that its *y*-values *increase 2 times as quickly* on the *xy*-plane.

» **Has no effect when *a* equals 1.** For example, $1f(x) = f(x)$, which leaves the function unchanged.

» **Compresses the function when *a* is less than 1.** For example, $\frac{1}{2}f(x)$ compresses the function, so that it *increases half as quickly* on the *xy*-plane.

Figure 12-6 shows these two stretch transformations applied to the quadratic parent function $f(x) = x^2$.

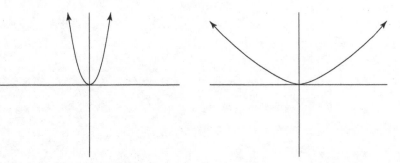

FIGURE 12-6:
Two positive
stretch-
compress
transforma-
tions of
$f(x) = x^2$.

If $f(x) = x^2$ and $p(x)$ is a transformation of $f(x)$ that includes the point $(-1, \frac{1}{4})$, then which of the following could be $p(x)$?

EXAMPLE

(A) $p(x) = 4x^2$

(B) $p(x) = 2x^2$

(C) $p(x) = \frac{1}{2}x^2$

(D) $p(x) = \frac{1}{4}x^2$

The function $f(x) = x^2$ includes the point $(-1,1)$. A stretch–compress transformation of this function that includes the point $(-1, \frac{1}{4})$ increases $\frac{1}{4}$ as quickly as the original function, so this is $p(x) = \frac{1}{4}x^2$. Therefore, Answer D is correct.

THE REFLECTION TRANSFORMATION WHEN VARIABLE *A* IS NEGATIVE

In Chapter 10, you discover that when a is negative, the transformation $af(x)$ reflects the function across the x-axis. Additionally, when a is *negative*, the transformation $af(x)$:

» **Flips *and* stretches the function when *a* is less than –1.** For example, $-2f(x)$ reflects the function across the x-axis and stretches the function, so that it *decreases 2 times as quickly* on the xy-plane.

» **Only flips the function when *a* equals –1.** For example, $-1f(x)$ reflects the function across the x-axis without stretching or compressing it.

» **Flips *and* compresses the function when $-1 < a < 0$.** For example, $-\frac{1}{2}f(x)$ reflects the function across the x-axis and compresses the function, so that it *decreases half as quickly* on the xy-plane.

Figure 12-7 shows these two reflection transformations applied to the quadratic parent function $f(x) = x^2$.

FIGURE 12-7:
Two negative stretch-compress-reflect transformations of $f(x) = x^2$.

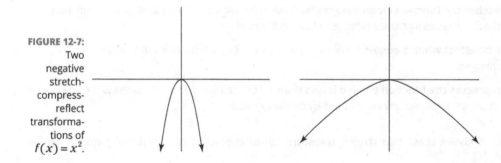

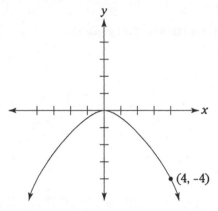

If $y = ax^2$ intersects both the origin and $(4,-4)$, then $a =$

(A) $-\dfrac{1}{8}$

(B) $-\dfrac{1}{4}$

(C) $\dfrac{1}{4}$

(D) $\dfrac{1}{8}$

The figure shows a parabola that's concave down, so a is negative, which rules out Answers C and D. Plug in 4 for x and -4 for y into $y = ax^2$ and solve for a:

$$-4 = a(4^2)$$
$$-4 = 16a$$
$$-\frac{1}{4} = a$$

Therefore, Answer B is correct.

Sketching a parabola from vertex form

Sketching a parabola on the xy-plane makes sense when you think of it as applying three transformations of the parent quadratic function $f(x) = x^2$: vertical, horizontal, and stretch/compress/reflect. Applying all three of these transformations to a quadratic function results in the vertex form $f(x) = a(x-h)^2 + k$.

To sketch a parabola from its vertex form quadratic equation, start at $(0,0)$, the anchor point for quadratic functions, and do the following:

➤ Use h and k to find the vertex of the function.

➤ Use a to account for the stretch/compress/reflect transformation of the function.

For example:

$$f(x) = 2(x+1)^2 - 1$$

Which of the following is the best graph of this function on the *xy*-plane?

(A)

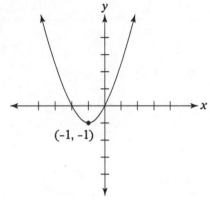

(-1, -1)

(B)

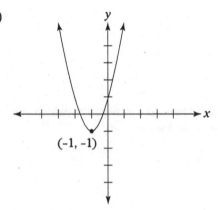

(-1, -1)

(C)

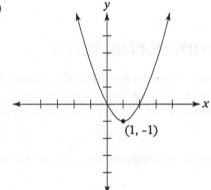

(1, -1)

(D)

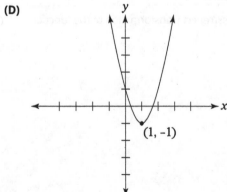

(1, -1)

Starting from $(0,0)$, the vertex of this function is displaced one unit to the left and one unit down, to $(-1,-1)$, which rules out Answers C and D. The function is then stretched by a factor of 2, which rules out Answer A, so Answer B is correct.

WARNING

When graphing a transformation with a reflection, the vertex stays in place as the function flips from concave up to concave down.

$$f(x) = -2(x-1)^2 + 4$$

EXAMPLE

Which of the following is the best graph of this function on the xy-plane?

(A)

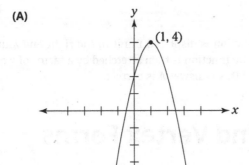

(B)

(C)

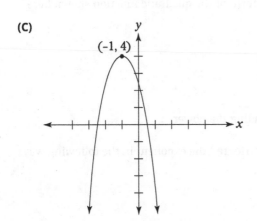

(D)

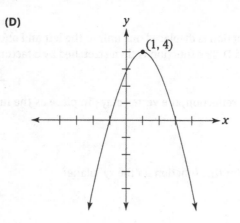

Starting from (0,0), the vertex of this function is displaced 1 unit to the right and 4 units up, to (1,4), which rules out Answers B and C. The function is then stretched by a factor of 2 and flipped to concave down, which rules out Answer D, so Answer A is correct.

Connecting Standard and Vertex Forms

So far in this chapter, you've worked with two different forms of the quadratic function.

>> Standard form: $f(x) = ax^2 + bx + c$

>> Vertex form: $f(x) = a(x - h)^2 + k$

In this section, you discover the connections between these two different forms.

Changing vertex form to standard form

To change a quadratic function from vertex form to standard form, just simplify the function. For example:

$$f(x) = 3(x - 4)^2 - 5$$

EXAMPLE

Which of the following is an equivalent form of the quadratic function shown here?

(A) $f(x) = 3x^2 + 4x - 5$

(B) $f(x) = 3x^2 - 4x - 5$

(C) $f(x) = 3x^2 - 24x + 33$

(D) $f(x) = 3x^2 - 24x + 43$

Begin simplifying the function by expanding the power.

A common student error is to try to "distribute" the exponent in the following way:

WARNING

$$(x - 4)^2 = x^2 - 16 \quad \text{WRONG!}$$

Here's the right way to do this:

$$(x - 4)^2 = (x - 4)(x - 4)$$

Thus, here's your first step when changing this function from vertex form to standard form:

$$3(x-4)^2 - 5$$
$$= 3(x-4)(x-4) - 5$$

From here, FOIL the contents of the parentheses:

$$= 3(x^2 - 8x + 16) - 5$$

Now, distribute the 3 and combine like terms:

$$= 3x^2 - 24x + 48 - 5$$
$$= 3x^2 - 24x + 43$$

Therefore, Answer D is correct.

Notice that when you change vertex form to standard form, the constant a remains the same. Keep this in mind when answering SAT questions.

Changing standard form to vertex form

One common method for changing quadratic equations from standard to vertex form is called *completing the square*. Some students like it, and others find it confusing.

In this section, I first show you how to perform this conversion without completing the square, using information about the standard form equation that you should already know. (To make sure, see the earlier section, "The Quadratic Function in Standard Form.")

Next, for good measure, I show you how to complete the square. In Chapter 15, you use this skill again when working with circles on the xy-plane.

Changing to vertex form *without* completing the square

The easier way to change standard form $(y = ax^2 + bx + c)$ quadratics to vertex form $(y = a(x-h)^2 + k)$ is based on the moves I show you earlier, in the section, "The Quadratic Function in Standard Form." Given a function $y = ax^2 + bx + c$, do the following:

1. Use $x = -\dfrac{b}{2a}$ to find the axis of symmetry, which is the h value in vertex form.
2. Plug this value into the function and solve for y, which is the k value in vertex form.
3. Using the a value from standard form, write $f(x) = a(x-h)^2 + k$.

For example:

EXAMPLE

Which of the following is an equivalent form of the quadratic function $y = -4x^2 + 8x - 9$?

(A) $y = -4(x-1)^2 + 5$

(B) $y = -4(x-1)^2 - 5$

(C) $y = -4(x+8)^2 - 9$

(D) $y = -4(x-8)^2 - 9$

In a pinch, you could answer this question by simplifying all four answers. For a faster way, begin by finding the axis of symmetry to find the value of h:

$$x = -\frac{b}{2a} = -\frac{8}{2(-4)} = -\frac{8}{-8} = 1$$

Plug this value into the function and simplify to find k:

$$y = -4(1)^2 + 8(1) - 9 = -4 + 8 - 9 = -5$$

Use the a value from the original function, along with these h and k values, to write the function:

$$y = a(x-h)^2 + k = -4(x-1)^2 - 5$$

Thus, Answer B is correct.

Completing the square

Completing the square is a way to change a standard form quadratic equation $y = ax^2 + bx + c$ into vertex form $f(x) = a(x-h)^2 + k$. You can also use it when working with circles on a graph, as I show you in Chapter 15.

UNDERSTANDING SQUARES OF BINOMIALS

To understand the basic idea behind completing the square, it's helpful to get comfortable with the squares of binomials. To do this, take a look at Table 12-1.

TABLE 12-1 Quadratic Trinomial Expressions as Equivalent Squares of Binomials

Trinomials with positive b-values	Trinomials with negative b-values
$x^2 + 2x + 1 = (x+1)^2$	$x^2 - 2x + 1 = (x-1)^2$
$x^2 + 4x + 4 = (x+2)^2$	$x^2 - 4x + 4 = (x-2)^2$
$x^2 + 6x + 9 = (x+3)^2$	$x^2 - 6x + 9 = (x-3)^2$
$x^2 + 8x + 16 = (x+4)^2$	$x^2 - 8x + 16 = (x-4)^2$
$x^2 + 10x + 25 = (x+5)^2$	$x^2 - 10x + 25 = (x-5)^2$

As you can see, certain quadratic expressions can be converted to an equivalent form as the square of a binomial. For example:

$$x^2 + 8x + 16 = (x+4)^2$$

But now, consider the expression $x^2 + 8x + 17$. While this expression isn't quite so friendly, you can rewrite it in this way:

$$x^2 + 8x + 17 = x^2 + 8x + 16 + 1$$

From here, you can factor the first three terms, rewriting the expression as follows:

$$(x+4)^2 + 1$$

If this basic idea makes sense to you, you're ready to try out completing the square. This method of working with quadratic trinomials can be confusing, and it's often taught in different ways by different teachers, which adds to the confusion.

So, if you're already completely comfortable with completing the square, feel free to skip over my explanation in the next section. Otherwise, try out the method I outline in the next two sections.

COMPLETING THE SQUARE FOR QUADRATICS WHEN *A* IS 1

The easier case of completing the square occurs when a is 1 — that is, when the quadratic you're starting with has the form $y = x^2 + bx + c$.

The trick to completing the square is to calculate a helpful value to introduce into your function using both addition and subtraction. This value is $\left(\frac{b}{2}\right)^2$. Here's an example:

EXAMPLE

$$y = x^2 + 10x + 23$$

Which of the following is the vertex form of the function?

(A) $y = (x+5)^2 + 2$

(B) $y = (x+5)^2 - 2$

(C) $y = (x-5)^2 + 2$

(D) $y = (x-5)^2 - 2$

To answer this question by completing the square, first substitute 10 for b into $\left(\frac{b}{2}\right)^2$:

$$\left(\frac{10}{2}\right)^2 = 5^2 = 25$$

Now, add and subtract this value into the function, just after the x-term (10x):

$$y = x^2 + 10x + 25 - 25 + 23$$

For the moment, this version of the function looks more complicated than when you started. But now, you can perform quadratic factoring on the first three terms of the polynomial $(x^2 + 10x + 25)$. I do this in two steps so you can see how it works:

$$y = (x+5)(x+5) - 25 + 23$$
$$y = (x+5)^2 - 25 + 23$$

To finish, simplify the last two terms:

$$y = (x+5)^2 - 2$$

So, Answer B is correct.

COMPLETING THE SQUARE WHEN *A* ISN'T 1

When the quadratic you're starting with has the form $y = ax^2 + bx + c$, with $a \neq 1$, completing the square requires a couple of additional steps. For example:

EXAMPLE

$$f(x) = 3x^2 - 18x + 17$$

If this function is equivalent to $f(x) = a(x-h)^2 + k$, then what is the value of k?

To complete the square, begin by factoring out the coefficient of the first term, which is 3, from the first two terms of the function $f(x) = 3x^2 - 18x + 17$:

$$f(x) = 3(x^2 - 6x) + 17$$

As you can see, this version of the function already looks a little more like vertex form than it did before. Now, calculate $\left(\frac{b}{2}\right)^2$, using -6 for b:

$$\left(\frac{-6}{2}\right)^2 = (-3)^2 = 9$$

Next, add and subtract the resulting value, 9, into the function *inside the parentheses*:

$$f(x) = 3(x^2 - 6x + 9 - 9) + 17$$

Now, here's where you have to be careful: The next step is to get the value -9 outside of the parentheses. To do this, distribute the value 3 by multiplying $-9 \times 3 = -27$:

$$f(x) = 3(x^2 - 6x + 9) - 27 + 17$$

At this point, you can perform quadratic factoring on the polynomial inside the parentheses $(x^2 - 6x + 9)$. As in the previous example, I do this in two steps:

$$f(x) = 3(x - 3)(x - 3) - 27 + 17$$
$$f(x) = 3(x - 3)^2 - 27 + 17$$

To finish, simplify the last two terms:

$$f(x) = 3(x - 3)^2 + 10$$

Thus, the value k is 10, so the correct answer is 10.

Finding the Roots of a Quadratic Function

In Chapter 11, I discuss the x-intercepts of a polynomial, which are the points where that function crosses the x-axis. I also alert you to the fact that the SAT uses the words *real roots*, *real solutions*, and *real zeros* relatively interchangeably to describe this same idea.

In this section, I clarify this distinction between real and non-real (complex) roots.

Distinguishing quadratic functions from quadratic equations

Students sometimes ask me, "What's the difference between a quadratic function and a quadratic equation?" Here's the answer:

>> A *quadratic function* has a variable on one side of the equals sign — typically, *y* or *f(x)*. For example, $y = x^2 - 7x + 10$ and $f(x) = -2x^2 + 5x - 17$ are both quadratic functions.

>> In a *quadratic equation*, 0 replaces the *y* or *f(x)* variable. For example, $0 = x^2 - 7x + 10$ and $0 = -2x^2 + 5x - 17$ are the two *related* quadratic equations for the functions shown here.

Getting clear on this point is helpful for finding the *roots* or *zeros* of a quadratic function — that is, the *solutions* of that function's related quadratic equation.

REMEMBER

To find the roots of the quadratic function $y = ax^2 + bx + c$, let $y = 0$ and solve. So, solving the quadratic equation $0 = ax^2 + bx + c$ gives you the roots of its related function.

In the next section, you use this understanding to find the x-intercepts (when they exist) of three quadratic functions.

Identifying quadratic functions with 2, 1, and no x-intercepts

Every quadratic function has 2, 1, or no x-intercepts. For example,

» $f(x) = x^2 - 1$ **has 2 x-intercepts:** 1 and –1

» $f(x) = x^2$ **has 1 x-intercept:** 0

» $f(x) = x^2 + 1$ **has no x-intercepts.**

Figure 12-8 shows you graphs of these three quadratic functions.

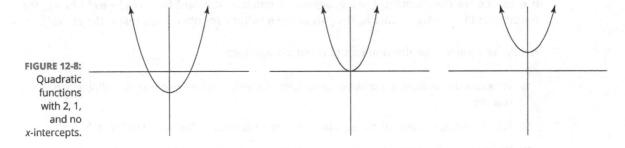

FIGURE 12-8:
Quadratic
functions
with 2, 1,
and no
x-intercepts.

As you can see, given the basic shape of a parabola, no quadratic equation can have more than two x-intercepts.

You can verify the x-intercepts of each of these quadratic functions by solving their related quadratic equation — that is, by setting it to 0 and solving for x.

$$x^2 - 1 = 0 \qquad\qquad x^2 = 0 \qquad\qquad x^2 + 1 = 0$$
$$x^2 = 1 \qquad\qquad \sqrt{x^2} = \pm\sqrt{0} \qquad\qquad x^2 = -1$$
$$\sqrt{x^2} = \pm\sqrt{1} \qquad\qquad x = 0 \qquad\qquad \sqrt{x^2} = \pm\sqrt{-1}$$
$$x = 1, -1 \qquad\qquad\qquad\qquad\qquad x = \pm i$$

As you can see, $x^2 - 1 = 0$ has two solutions, 1 and –1, so it has x-intercepts at both of these points. Similarly, $x^2 = 0$ has only one solution, 0, so it has an x-intercept only at this point.

As for $x^2 + 1 = 0$, it has no real solutions, so it has no x-intercepts. However, it has two solutions that include the imaginary number $i = \sqrt{-1}$ — that is, it has no real solutions, but two *complex solutions* — a pair of solutions of the form $a \pm bi$.

In Table 12-2, I organize this information about these three types of quadratic functions. As you can see, the terms *real roots*, *real solutions*, and *real zeros* are all synonymous with x-intercepts — that is, points on the xy-plane where the graph intersects the x-axis. Similarly, the terms *complex roots*, *complex solutions*, and *complex zeros* all refer to solutions to a quadratic equation that include the imaginary number i.

TIP

When an SAT question asks for the *roots*, *solutions*, or *zeros* of an equation without specifying which type — real or complex — you can assume that they're referring to *real* values.

TABLE 12-2 Distinguishing *x*-Intercepts from Complex Roots

x-Intercepts	Complex Solutions
Also called: real roots, real solutions, real zeros	Also called: complex roots, complex zeros
2	0
1	0
0	2

In this book, I try to stick to the unambiguous term *x*-intercepts to refer to real solutions to a quadratic or other polynomial equation.

Using the discriminant to count *x*-intercepts

A relatively simple way to count the *x*-intercepts of a quadratic function without actually solving it is to find the discriminant. For a quadratic equation in standard form $f(x) = ax^2 + bx + c$, the discriminant is $b^2 - 4ac$ — that is, the part of the quadratic formula that's inside the radical.

Here's how you use the discriminant to count *x*-intercepts:

>> When the discriminant is a positive number, the equation has two *x*-intercepts — that is, two real roots.

>> When the discriminant is 0, the equation has one *x*-intercept — that is, one real root.

>> When the discriminant is a negative number, the equation has no *x*-intercepts — that is, two complex roots — one conjugate pair of the form $a \pm bi$.

EXAMPLE

How many *x*-intercepts does the function $f(x) = 3x^2 + 2x - 5$ have when graphed on the *xy*-plane?

(A) 0

(B) 1

(C) 2

(D) 3

The function is quadratic, so it has a maximum of two *x*-intercepts, which means you can rule out Answer D. To find out how many *x*-intercepts it has, evaluate the discriminant $b^2 - 4ac$, plugging in 3 for *a*, 2 for *b*, and −5 for *c*:

$$b^2 - 4ac = 2^2 - 4(3)(-5) = 4 + 60 = 64$$

The result is positive, so the function has two *x*-intercepts; therefore, Answer C is correct.

EXAMPLE

$$f(x) = 2x^2 + 5x + 6$$

This function has

(A) One real root

(B) Two real roots

(C) One real root and one complex root

(D) Two complex roots

To begin, a quadratic function never has one real root and one complex root, so you can rule out Answer C. To find the answer, plug in 2 for a, 5 for b, and 6 for c into the discriminant $b^2 - 4ac$ and evaluate:

$$5^2 - 4(2)(6) = 25 - 48 = -23$$

The result is negative, so the function has two complex roots; therefore, Answer D is correct.

Finding the roots of a quadratic function

In Chapter 3, you discover three ways to solve quadratic equations.

>> **Isolating x:** This method works only for two-term quadratic equations, such as $x^2 - 9 = 0$ and $x^2 + 5x = 0$, but it's very quick and simple.

>> **Factoring:** This method works only for quadratic equations that are factorable, such as $x^2 + 4x + 3 = 0$ and $2x^2 + 3x - 5 = 0$, but in many cases, you'll find it to be the quickest way to solve a quadratic.

>> **Using the quadratic equation:** This method works for all equations, but it's time consuming and requires care to get the correct answer.

Check out that chapter now if you need a review of these methods for finding the solutions of a quadratic equation. I use them here to answer SAT questions where the best way to find the roots of a quadratic function is by solving its related quadratic equation.

Here's an example of a two-term quadratic function that's relatively easy to solve (flip to Chapter 3 if you're not quite sure how this method of solving works):

$$f(x) = 2x^2 - 9x$$

If $k > 0$ is a root of this function, what is the value of k?

EXAMPLE

As always when finding the roots of a function, begin by setting it to 0:

$$2x^2 - 9x = 0$$

To solve, factor out an x on the left side of the equation:

$$x(2x - 9) = 0$$

Now, set both factors equal to x and solve the resulting equations separately:

$$x = 0 \qquad\qquad\qquad 2x - 9 = 0$$
$$2x = 9$$
$$x = \frac{9}{2}$$

Thus, because $k > 0$, the answer is $\frac{9}{2}$, which you can also grid in as 4.5.

Here's an example of a factorable quadratic that's also not too difficult to solve (see Chapter 3 if you need a refresher on factoring quadratics):

$$f(x) = x^2 - 7x - 18$$

If a and b are the zeros of this function, what is the value of $|a| + |b|$?

EXAMPLE

Begin by setting the function to 0:

$$x^2 - 7x + 18 = 0$$

This quadratic expression is factorable:

$$(x-9)(x+2) = 0$$

Now, set each factor equal to 0 and solve separately:

$$x - 9 = 0 \qquad\qquad x + 2 = 0$$
$$x = 9 \qquad\qquad x = -2$$

Therefore, the values of p and q are 9 and -2, so $|a| + |b| = |9| + |-2| = 9 + 2 = 11$, making the answer 11.

Here's a question that looks difficult at first, but turns out to be factorable:

EXAMPLE

If $n > 0$, what value of n is a zero of the function $f(x) = 3x^2 + 9x - 12$?

Begin by setting the function to 0:

$$0 = 3x^2 + 9x - 12$$

Dividing both sides of this equation by 3 leaves the left side unchanged, but allows you to factor more easily:

$$0 = x^2 + 3x - 4$$
$$0 = (x+4)(x-1)$$

Thus, $x = -4$ or $x = 1$, so the zeros of the equation are -4 and 1. Because the question specifies that $n > 0$, the correct answer is 1.

Finally, here's one more example that requires the quadratic formula to solve it (see Chapter 3 for a refresher on the quadratic formula):

EXAMPLE

If p and q are the zeros of the function $f(x) = x^2 - 5x - 8$, then what is the sum of p and q?

(A) 5

(B) -5

(C) $5 + \sqrt{57}$

(D) $5 - \sqrt{57}$

This time, the function isn't factorable, so plug in $a = 1$, $b = -5$, and $c = -8$ into the quadratic formula and solve:

$$
\begin{aligned}
x &= \frac{-b \pm \sqrt{b^2 - 4ac}}{2a} \\
&= \frac{-(-5) \pm \sqrt{(-5)^2 - 4(1)(-8)}}{2(1)} \\
&= \frac{5 \pm \sqrt{25 + 32}}{2} \\
&= \frac{5 \pm \sqrt{57}}{2}
\end{aligned}
$$

Thus $x = \dfrac{5 + \sqrt{57}}{2}$ and $x = \dfrac{5 - \sqrt{57}}{2}$. Add these two values together as follows:

$$\frac{5 + \sqrt{57}}{2} + \frac{5 - \sqrt{57}}{2} = \frac{5 + \sqrt{57} + 5 - \sqrt{57}}{2} = \frac{10}{2} = 5$$

Therefore, Answer A is correct.

Using quadratic equations to solve word problems about projectiles

A common type of SAT question uses a quadratic function $f(x) = ax^2 + bx + c$ to model the motion of a projectile — that is, an object projected into the air and allowed to fall back to Earth. In these problems, each of the three values a, b, and c has a specific role to play in the model.

>> **a is acceleration, the force of gravity.** On Earth, $a = -16$ when you're measuring velocity in feet per second — that is, in English units, which are the units that the SAT tends to use. (If you take physics, where the units used are typically meters per second, you may be used to the value –4.9, but in any case, an SAT question will normally provide you with this value rather than expect you to know it.)

>> **b is the initial velocity of the projectile.** For example, if a soccer ball is thrown into the air at 12 feet per second, $b = 12$.

>> **c is the initial height of the projectile.** For example, if a soccer ball is thrown from the top of a 40-foot platform, $c = 40$.

Here's how you can put this information to work on an SAT question.

While standing on a platform raised 40 feet above the ground, Jacob kicked a soccer ball in the air at an initial upward velocity of 12 feet per second. The height of the ball, in feet, x seconds after Jacob kicked it, was $h(x) = -16x^2 + 12x + 40$. How much time, in seconds, did the ball remain in the air before it landed on the ground?

(A) –1.25

(B) 0.375

(C) 1.25

(D) 2.0

Before proceeding to the question, draw a quick sketch of the graph. Because $a = -16$ is negative, the graph is concave down, which makes sense when modeling the flight of a projectile! Because $c = 40$, the y-intercept is at (0,40), indicating where Jacob was standing when he kicked the ball.

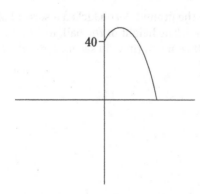

As you can see, the time when the ball landed corresponds with an x-intercept of the graph. To find this value, set the function to 0:

$$0 = -16x^2 + 12x + 40$$

To make this equation easier to work with, divide both sides by -4:

$$0 = 4x^2 - 3x - 10$$

This equation is factorable, so solve by factoring (see Chapter 3 for details on factoring quadratic equations):

$$0 = 4x^2 + 5x - 8x - 10$$
$$0 = x(4x + 5) - 2(4x + 5)$$
$$0 = (x - 2)(4x + 5)$$

To complete the problem, you can split this equation into two equations and solve each one separately:

$$\begin{array}{ccc} 0 = x - 2 & & 0 = 4x + 5 \\ 2 = x & \text{OR} & -5 = 4x \\ & & -1.25 = x \end{array}$$

To interpret these results, I fill these values into the sketch shown here:

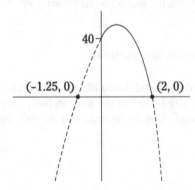

Note that I've added a dotted line before time 0, to show the part of the graph that has no real-world significance (unless Jacob owns a time machine, so he could travel back in time to 1.25 seconds before he kicked it!). Thus, the ball landed 2 seconds after Jacob kicked it, so Answer D is correct.

Suppose the question had been different, as follows:

EXAMPLE

While standing on a platform raised 40 feet above the ground, Jacob kicked a soccer ball in the air at an initial upward velocity of 12 feet per second. The height of the ball, in feet, x seconds after Jacob kicked it, was $h(x) = -16x^2 + 12x + 40$. How much time, in seconds, did the ball require to reach its maximum height?

(A) -1.25

(B) 0.375

(C) 1.25

(D) 2.0

To answer this question, find the axis of symmetry:

$$x = -\frac{b}{2a} = -\frac{12}{2(-16)} = -\frac{12}{-32} = \frac{3}{8} = 0.375$$

Thus, the ball reached its maximum height 0.375 second after Jacob kicked it, so Answer B is correct.

Here's one final question:

EXAMPLE

While standing on a platform raised 40 feet above the ground, Jacob kicked a soccer ball in the air at an initial upward velocity of 12 feet per second. The height of the ball, in feet, x seconds after Jacob kicked it, was $h(x) = -16x^2 + 12x + 40$. What was the maximum height of the ball, as measured in feet above the ground?

(A) 41.5

(B) 42.25

(C) 46.75

(D) 50.5

To answer this question, plug the value of the axis of symmetry (which you found to answer the previous question) into the function and solve to find the y-value of the vertex:

$$-16(0.375)^2 + 12(0.375) + 40 = -2.25 + 4.5 + 40 = 42.25$$

Thus, the ball reached a maximum height of 42.25 feet above the ground, so Answer B is correct.

I show all the information I gathered when answering these three questions in this final sketch of the graph.

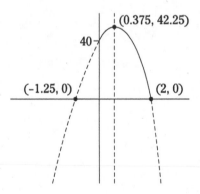

Chapter **13**

Exponential and Radical Equations

E xponential and radical equations are among the more advanced algebra equations that you'll see on the SAT. The good news here is that, although these equations are advanced, the types of problems that are on the SAT are somewhat basic. So in many cases, you only need to know a handful of concepts to do well solving these problems.

In this chapter, I start you off with a set of exponential identities that allow you to simplify expressions that include exponents. Then, you use these identities to solve exponential equations. Next, you work with exponential growth and decay functions, and use them to model a variety of real-world phenomena to answer word problems.

After that, the focus is on radicals. First, you review a variety of basic facts about radicals, including applying basic operations, simplifying, and rationalizing the denominator. Next, you solve radical equations, especially those that may produce extraneous solutions. To finish up, you graph radical functions on the xy-plane.

Solving Exponential Equations

An exponential equation includes at least one variable, such as x, in an exponent. In this section, I first show you how to simplify exponential expressions using a set of identities that you may already be familiar with. After that, I show you how to use these identities to solve the types of exponential equations that you may find on the SAT.

Using exponential identities

An identity is an equation that works for an infinite number of combinations of variables, such as $5x + 2x = 7x$.

On the SAT, it's helpful to know how to use identities that include exponents to simplify exponential expressions and solve exponential equations. Table 13-1 lists five important identities.

TABLE 13-1 **Identities for Simplifying Exponential Expressions**

Operation	Identity	Example
Multiplication	$x^a x^b = x^{a+b}$	$x^3 x^4 = x^{3+4} = x^7$
Division	$\dfrac{x^a}{x^b} = x^{a-b}$	$\dfrac{x^9}{x^5} = x^{9-5} = x^4$
Exponents of exponents	$(x^a)^b = x^{ab}$	$(x^2)^6 = x^{2 \cdot 6} = x^{12}$
Negative exponents	$x^{-a} = \dfrac{1}{x^a}$ and $\dfrac{1}{x^{-a}} = x^a$	$x^{-5} = \dfrac{1}{x^5}$ and $\dfrac{1}{x^{-8}} = x^8$
Fractional exponents	$x^{\frac{a}{b}} = \sqrt[b]{x^a} = (\sqrt[b]{x})^a$	$x^{\frac{2}{3}} = \sqrt[3]{x^2} = (\sqrt[3]{x})^2$

Some SAT math questions focus directly on these identities, asking you to simplify or rewrite a given exponential expression. For example:

EXAMPLE

$$\frac{x^7 y^{-4}}{x^{-5} y^{-3}}$$

This expression is equivalent to which of the following?

(A) $\dfrac{x^2}{y^7}$

(B) $\dfrac{y}{x^2}$

(C) $\dfrac{x^{12}}{y}$

(D) $\dfrac{y^7}{x^{12}}$

Begin by using the two negative exponent identities to move the three negative exponents in this expression to the opposite side of the fraction bar, flipping their signs from negative to positive:

$$\frac{x^7 y^{-4}}{x^{-5} y^{-3}} = \frac{x^7 x^5 y^3}{y^4}$$

Now, use the multiplication identity to simplify the x-values and the division identity to simplify the y-values:

$$= x^{12} y^{-1}$$

Complete the problem by applying the negative exponent identity to the y-value:

$$\frac{x^{12}}{y}$$

Thus, Answer C is correct.

As another example:

EXAMPLE

$$(x^2)^{\frac{3}{4}}(y^4)^{\frac{1}{8}}$$

Which of the following is equivalent to this expression?

(A) $\sqrt{x^3}\sqrt{y}$

(B) $\sqrt{\dfrac{x^3}{y}}$

(C) $\sqrt[3]{x^2}\sqrt{y}$

(D) $\dfrac{\sqrt[3]{x^2}}{\sqrt{y}}$

To begin, using the identity for exponents of exponents, multiply the exponents for both x and y. This involves fraction multiplication, which I do in two steps:

$$(x^2)^{\frac{3}{4}}(y^4)^{\frac{1}{8}} = x^{\frac{6}{4}}y^{\frac{4}{8}} = x^{\frac{3}{2}}y^{\frac{1}{2}}$$

Now, use the identity for fractional exponents:

$$\sqrt{x^3}\sqrt{y}$$

Thus, Answer A is correct.

You can also use these identities to simplify expressions with numerical bases. For example:

EXAMPLE

What is the integer value of $81^{\frac{3}{4}}3^{-2}$?

To begin, rewrite $81^{\frac{3}{4}}3^{-2}$ using the exponential identities for fractional and negative exponents:

$$81^{\frac{3}{4}}3^{-2} = (\sqrt[4]{81})^3\,\frac{1}{3^2} = \frac{(\sqrt[4]{81})^3}{9}$$

To simplify this value further, note that $\sqrt[4]{81} = 3$, because $3^4 = 81$:

$$= \frac{3^3}{9} = \frac{27}{9} = 3$$

Therefore, the correct answer is 3.

Solving exponential equations

An exponential equation includes at least one term that has a variable such as x in its exponent. An SAT math question may ask you to demonstrate your understanding of the exponential identities to solve an exponential equation.

Here's a general strategy for solving the types of exponential equations that appear on the SAT:

1. **Using exponential identities, rewrite one or both sides of the equation in terms of a common base — that is, using the same base of the exponent for both sides.** In most cases, this will be a relatively small integer.

2. **Drop the bases and set the two exponents equal to each other.**

3. **Solve for the variable.**

$$2^{x+4} = 8^x$$

What is the value of x in this equation?

To begin, notice that because $2^3 = 8$, you can rewrite 8^x in the equation using a base of 2 as follows:

$$2^{x+4} = (2^3)^x$$

As you can see, this change to the right side of the equation requires parentheses. But now, you can use the identity $(x^a)^b = x^{ab}$ to rewrite this side of the equation without parentheses:

$$2^{x+4} = 2^{3x}$$

Now, because both sides of the equation have a common base of 2, you can drop this base and set the two exponents equal to each other:

$$x + 4 = 3x$$

This equation is easy to solve:

$$4 = 2x$$
$$2 = x$$

Therefore, the correct answer is 2.

Here's another example:

If $25^{x-7} = (\frac{1}{5})^{2x}$, what is the value of x?

In this case, you can rewrite the base of 25 as 5^2 and the base of $\frac{1}{5}$ as 5^{-1}. Be sure to use parentheses in both cases:

$$(5^2)^{x-7} = (5^{-1})^{2x}$$

Again, use the identity $(x^a)^b = x^{ab}$ to rewrite both sides of the equation:

$$5^{2(x-7)} = 5^{-1(2x)}$$

Now that both sides of the equation have a common base of 5, you can drop the bases, set the exponents equal, and solve for x:

$$2(x - 7) = -1(2x)$$
$$2x - 14 = -2x$$
$$-14 = -4x$$
$$\frac{7}{2} = x$$

Therefore, the correct answer is $\frac{7}{2}$, which is also equivalent to 3.5.

Here's one final problem where you need to rewrite a root in terms of a fractional exponent:

$$\sqrt[3]{2} = 4^{x-1}$$

In this equation, what is the value of x?

To begin, rewrite the cube root $\sqrt[3]{2}$ using the exponent $\frac{1}{3}$, and the base of 4 as 2^2:

$$2^{\frac{1}{3}} = (2^2)^{x-1}$$

Simplify using $(x^a)^b = x^{ab}$:

$$2^{\frac{1}{3}} = 2^{2(x-1)}$$

To complete the problem, drop the bases, set the exponents equal, and solve for x:

$$\frac{1}{3} = 2(x-1)$$

$$\frac{1}{3} = 2x - 2$$

$$\frac{1}{3} + 2 = 2x$$

$$\frac{7}{3} = 2x$$

$$\frac{7}{6} = x$$

Therefore, the correct answer is $\frac{7}{6}$, which you can also write as 1.16 or 1.17.

Answering SAT Math Questions Using Exponential Functions

The exponential function $f(x) = ab^x$ can be used to model a variety of real-world situations. It divides into two main groups, depending on the value of b.

>> **Exponential growth functions:** When $b > 1$, the exponential function always increases as x increases.

>> **Exponential decay functions:** When $0 < b < 1$, the exponential function always decreases as x increases.

In this section, you work with both of these types of exponential functions.

Understanding exponential growth

When $b > 1$, the exponential function $f(x) = ab^x$ is always increasing as x increases. In this equation:

>> **The constant a is the y-intercept,** the starting value for the function.

>> **The constant b is the multiplier,** the amount by which the value is multiplied as x increases by 1.

Table 13-2 presents a few examples to help you understand exponential growth functions.

In the first row, $a = 1$, so it appears to drop out of the function, and $b = 2$. As a result, the function starts at 1 and is multiplied by 2 as x increases by 1.

In the second row, $a = 100$, so this value is multiplied by 2. The third and fourth rows have the same starting value of 100, but show percent increases of 10% and 4%, reflected in the b-values of 1.1 and 1.04.

TABLE 13-2 Four Exponential Growth Functions

$f(x) = ab^x$	$f(0)$	$f(1)$	$f(2)$	$f(3)$
$f(x) = 2^x$	1	2	4	8
$f(x) = 100(2)^x$	100	200	400	800
$f(x) = 100(1.1)^x$	100	110	121	133.1
$f(x) = 100(1.04)^x$	100	104	104.16	112.4864

As you can see, as x increases, an exponential growth function always increases at an accelerating rate. So, in each case, the change from $f(0)$ to $f(1)$ is less than the change from $f(1)$ to $f(2)$, and this trend continues infinitely.

This feature distinguishes an exponential growth function from an increasing linear function, which has a fixed rate of growth (its slope).

Some SAT math questions test your understanding of this important difference between linear and exponential growth by presenting you with a word problem. For example:

EXAMPLE

Four years ago, Jason earned $3,000 doing private catering jobs. Every year since then, he's doubled his catering income. Which of the following types of functions best describes the yearly change in Jason's income?

(A) Linear increase

(B) Linear decrease

(C) Exponential growth

(D) Exponential decay

Because Jason's income is increasing, you can rule out Answers B and D. According to the question, his catering income in the last four years has been $3,000, $6,000, $12,000, and $24,000. This rate of increase is, itself, increasing, which indicates exponential growth. Therefore, Answer C is correct.

Another type of SAT math question may give you a table of values and ask you to identify the type of function:

EXAMPLE

x	$f(x)$
0	7
1	21
2	63
3	189

This table shows four values of $f(x)$. Which of the following best describes the function?

(A) Linear increase

(B) Linear decrease

(C) Exponential growth

(D) Exponential decay

The values in the table are increasing, which rules out Answers B and D.

The starting value of the function is 7, and it is increasing by a factor of 3. Therefore, this is an exponential growth function, so Answer C is correct.

A moderately difficult SAT math question might ask you to identify the meaning of one or more values in an exponential growth function. For example:

EXAMPLE

A bacteria population is modeled by the function $f(h) = 12{,}000(1.5)^h$, where h equals the number of hours after 12:00 p.m. on the first day of an experiment. In this function, what does the value 12,000 signify?

(A) The number of bacteria at time 0.

(B) The increase in population only in the first hour after time 0.

(C) The increase in population during any hour in the experiment.

(D) The time at which the experiment begins.

The function is exponential, so 12,000 indicates the number of bacteria at the start of the experiment. You can verify this by calculating $f(0)$:

$$f(0) = 12{,}000(1.5)^0 = 12{,}000(1) = 12{,}000$$

Therefore, Answer A is correct.

A more difficult question might require you to set up and solve an exponential growth function given a set of parameters. For example:

EXAMPLE

Jasmyn invested \$2,200 in an annuity that has an annual growth rate of 3.5%. To the nearest whole dollar, what will the annuity be worth in 3 years?

Use the exponential growth function $f(x) = a(b)^x$ to model the value of this annuity x years after the investment begins. The starting value a is 2,200. To find the value of b, remember that a percent increase of 3.5% uses a multiplier of 103.5%, which is equivalent to the decimal 1.035. (For more on working with percent increase, flip to Chapter 7.)

So use the following function to model the annuity:

$$f(x) = 2200(1.035)^x$$

To answer the question, find the value of the function when $x = 3$. Typically, an SAT question wouldn't ask you to do this without a calculator, so be sure you know how to do a calculation that uses an exponent of 3:

$$f(3) = 2200(1.035)^3 \approx 2439.17$$

The question asks you to round your result to the nearest whole dollar, so the correct answer is 2,439.

A hard exponential problem might require you to discover an exponential growth function from relatively obscure information. For example:

EXAMPLE

$$f(x) = kn^x$$

Given this function, if $f(3) = 96$ and $f(5) = 1536$, what is the value of $f(4)$?

To begin solving this problem, write out the two equations in terms of k and n:

$$kn^3 = 96 \qquad kn^5 = 1536$$

Next, substitute 96 for kn^3 into the equation $kn^5 = 1536$. I do this in two steps so you can see what's happening:

$$kn^5 = 1536$$
$$kn^3 n^2 = 1536$$
$$96n^2 = 1536$$

Now, solve for n:

$$n^2 = 16$$
$$n = 4$$

Plug this value back into either of the original equations and solve for k:

$$kn^3 = 96$$
$$k(4)^3 = 96$$
$$64k = 96$$
$$k = 1.5$$

Thus, the original function is $f(x) = 1.5(4)^x$, so:

$$f(4) = 1.5(4)^4 = 1.5(256) = 384$$

Therefore, the correct answer is 384.

Working with exponential decay

When $0 < b < 1$, the exponential function $f(x) = ab^x$ is always decreasing as x increases. As in the exponential growth function:

>> **The constant a is the y-intercept,** the starting value for the function.

>> **The constant b is the multiplier,** the amount by which the value is multiplied as x increases by 1.

Table 13-3 presents a few examples of exponential decay functions.

TABLE 13-3 **Four Exponential Decay Functions**

$f(x) = ab^x$	$f(0)$	$f(1)$	$f(2)$	$f(3)$
$f(x) = 0.5^x$	1	0.5	0.25	0.125
$f(x) = 100(0.5)^x$	100	50	25	12.5
$f(x) = 100(0.75)^x$	100	75	56.25	42.1875
$f(x) = 100(0.9)^x$	100	90	81	72.9

In the first row, $a = 1$, so it appears to drop out of the function, and $b = 0.5$. As a result, the function starts at 1 and is multiplied by 0.5 as x increases by 1.

In the second row, $a = 100$, so this value is multiplied by 0.5. The third and fourth rows have the same starting value of 100, but show percent decreases of 25% and 10%, reflected in the b-values of 0.75 and 0.9.

As you can see, as x increases, an exponential decay function always decreases at a decreasing rate — that is, the value of the function approaches a horizontal asymptote. So, in each case, the change from $f(0)$ to $f(1)$ is greater than the change from $f(1)$ to $f(2)$, and this trend continues infinitely.

As with exponential growth, exponential decay differs from linear decrease, which has a fixed rate of decrease (its slope).

Here's an SAT math question that tests your understanding of this key difference between linear decrease and exponential decay:

EXAMPLE

Caitlyn's teacher assigned the class a novel to read over spring break. She read 160 pages the first day, but on each of the five days after that, she read only half as many pages as the day before. Which of the following types of functions would best model Caitlyn's reading for these six days?

(A) Linear increase

(B) Linear decrease

(C) Exponential growth

(D) Exponential decay

According to the question, Caitlyn read for six days, covering 160, 80, 40, 20, 10, and 5 pages per day, respectively. This shows a decreasing trend, which eliminates Answers A and C. If this were a linear relationship, the decrease in pages would be constant each day, but this is not the case, ruling out Answer B. Instead, her progress through the book shows exponential decay, so Answer D is correct.

A slightly more difficult question might ask you to identify an exponential decay function that models a specific situation. For example:

EXAMPLE

Caitlyn's teacher assigned the class a novel to read over spring break. She read 160 pages the first day, but on each of the five days after that, she read only half as many pages as the day before. If $0 \le x \le 5$, with x representing the whole number of days after Caitlyn started reading the novel, which of the following functions best models Caitlyn's reading for these six days?

(A) $f(x) = 160x^{0.5}$

(B) $f(x) = 0.5x^{160}$

(C) $f(x) = 0.5(160)^x$

(D) $f(x) = 160(0.5)^x$

Caitlyn's reading is best modeled by an exponential decay function. Answers A and B are ruled out because they're not exponential functions. The function in Answer C has a base that's greater than 1, so it's an exponential growth function, and is therefore ruled out. Answer D has a y-intercept of 160, and as x increases by 1, this value is halved. Thus, Answer D is correct.

In some cases, an SAT math question will ask you to calculate the value of an exponential function for a specific input value. For example:

EXAMPLE

$$f(t) = d(0.6)^t$$

When trading used computers online, Eileen uses this exponential function to estimate the value of a computer that originally sold for d dollars when it was new t years ago. To the nearest $100, how much value would Eileen place on a computer whose selling price out of the box 3 years ago was $1,400?

To solve this problem, plug in 1400 for d and 3 for t:

$$f(3) = 1400(0.6)^3 = 302.4 \approx 300$$

Thus, Eileen would value this computer at $302.40, which is approximately $300.

As with exponential growth functions, an SAT math question may ask you to model a real-world scenario using an exponential decay function. For example:

EXAMPLE

DynaCore stock currently sells for k per share. However, a stock analyst is predicting that the stock will lose 6% of its value for each of the first six months after the new year. Assuming that this prediction is accurate, and $m \le 6$, which of the following functions should correctly predict the value of DynaCore stock m months after the new year?

(A) $f(m) = k(0.06)^m$

(B) $f(m) = k(0.6)^m$

(C) $f(m) = k(0.94)^m$

(D) $f(m) = k(1.06)^m$

The analyst is predicting a 6% decrease in value per month. You can calculate this percent decrease by taking 94% of the value of the stock each month (see Chapter 7 for more on calculating percent decrease). This percentage is equivalent to the decimal 0.94, so this value is the base of the exponent. Therefore, Answer C is correct.

A more difficult SAT math question may provide you with an exponential decay function and ask you to solve for a constant inside that function. For example:

EXAMPLE

DynaCore stock currently sells for k per share. However, a stock analyst is predicting that the stock will lose 6% of its value for each of the first six months after the new year. If the price of DynaCore stock is predicted to fall to $50 six months after the new year, which of the following is closest to the current value of k?

(A) $60

(B) $64

(C) $68

(D) $72

The exponential decay function $f(m) = k(0.94)^m$ models the predicted value of DynaCore stock. According to this model, the value of the stock will fall to $50 in six months, so:

$$f(6) = k(0.94)^6 = 50$$

To solve this equation for k, begin by evaluating $(0.94)^6 \approx 0.69$, so:

$$0.69k = 50$$

To complete the problem, divide both sides of this equation by 0.69:

$$\frac{0.69k}{0.69} = \frac{50}{0.69}$$

$$k \approx 72.46$$

Therefore, Answer D is correct.

Graphing Exponential Functions

Figure 13-1 shows the shapes of the two main categories of exponential functions.

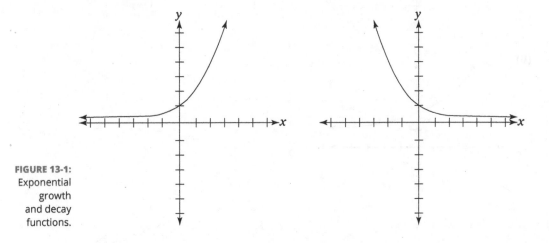

A relatively straightforward SAT math question may present you with a graph and ask you to classify the function shown. For example:

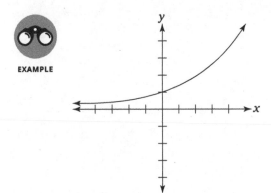

EXAMPLE

Which of the following best describes this function?

(A) Linear increase

(B) Linear decrease

(C) Exponential growth

(D) Exponential decay

The graph is increasing in a non-linear way, so the only possible answer among the four choices is Answer C.

A more difficult problem may present you with an exponential function and ask you to identify how values in that function affect its graph on the *xy*-plane. For example:

Which of the following is the best sketch of the function $f(x) = 2(3)^x$?

(A)

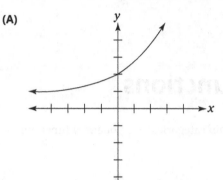

(B)

(C)

(D)

This function takes the form $f(x) = ab^x$, so the *y*-intercept *a* is 2, which rules out Answers C and D. The base *b* of this function is 3, which is greater than 1, so this is an exponential growth function, which rules out Answer B. Therefore, even though the sketch isn't perfect, Answer A is correct.

Recall that the function $f(x) = 2^x$ stands in as the exponential parent function (because the constant e is usually considered outside the scope of the SAT). The anchor point for this function is $(0,1)$, which you can use to help place vertical and horizontal transformations of the function. For example:

EXAMPLE

If $f(x) = 2^x$, which of the following is the graph of $f(x+2)-1$ that best describes this function?

(A)

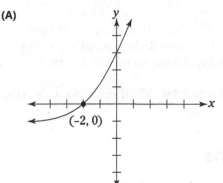

$(-2, 0)$

(B)

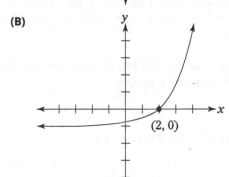

$(2, 0)$

(C)

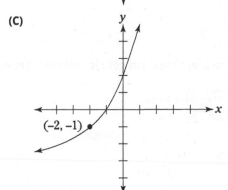

$(-2, -1)$

(D)

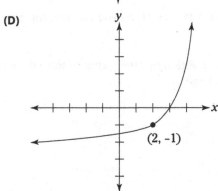

$(2, -1)$

The function $f(x) = 2^x$ has its anchor point at $(0,1)$. The transformation $f(x+2)-1$ moves the anchor point left 2 units and down 1 unit, to $(-2,0)$. Therefore, Answer A is correct.

Radical Equations and Functions

Radicals — also called square roots, such as $\sqrt{2}$ — arise from reversing the process of squaring a number. In Chapter 2, I give you a refresher on simplifying radical expressions, and applying operations such as addition, subtraction, multiplication, and division to radicals.

In this section, you build on this information to solve radical equations. I also show you how to graph radical functions on the xy-plane.

Solving radical equations

Radical equations can be a little confusing because normal solving methods can produce *extraneous* solutions — that is, solutions that appear to be correct but which, in fact, fail to solve the original equation.

To help you understand why this happens, and how to avoid it, I'll start here with an SAT math question that offers possible solutions that you can try plugging in:

EXAMPLE

$$\sqrt{4x+24} - 3 = x$$

Which of the following is the complete solution set for this equation?

(A) $\{3\}$

(B) $\{-5\}$

(C) $\{3, -5\}$

(D) The equation has no real solutions.

The question offers only two possible solutions, so plug them both in and see what you get:

$$\sqrt{4(3)+24} - 3 = 3 \qquad\qquad \sqrt{4(-5)+24} - 3 = -5$$
$$\sqrt{12+24} - 3 = 3 \qquad\qquad \sqrt{-20+24} - 3 = -5$$
$$\sqrt{36} - 3 = 3 \quad \text{CORRECT!} \qquad \sqrt{4} - 3 = -5 \quad \text{WRONG!}$$
$$6 - 3 = 3 \qquad\qquad\qquad\qquad 2 - 3 = -5$$
$$3 = 3 \qquad\qquad\qquad\qquad\qquad 1 = -5$$

Thus, 3 is a solution and −5 isn't, so Answer A is correct. As you can see, the plug-in solution presents an unambiguous answer.

But here's what happens if you approach the question by attempting to solve the equation using the method you were most likely taught in school:

$$\sqrt{4x+24} - 3 = x$$
$$\sqrt{4x+24} = x+3$$
$$\left(\sqrt{4x+24}\right)^2 = (x+3)^2$$
$$4x+24 = x^2+6x+9$$
$$0 = x^2+2x-15$$
$$0 = (x-3)(x+5)$$
$$x = 3, -5$$

As you can see, this method of solving appears to produce two solutions, one of which (−5) is extraneous.

To understand why this happens, take another look at this line of the original plug-in solution for −5:

$$\sqrt{4} - 3 = -5$$

This equation is incorrect because $\sqrt{4} = 2$. But consider that the value −2 also has the property that when multiplied by itself, the result is 4. So, while the symbol $\sqrt{4}$ unambiguously equals 2, the "ghost" value of −2 occasionally gets into the mix when the solution to an equation involves squaring a square root.

Graphing radical functions

The parent radical function is:

$$f(x) = \sqrt{x}$$

This function is shown in Figure 13-2. Note that this function is slightly different from the inverse of the function $f(x) = x^2$, because it includes only positive output values.

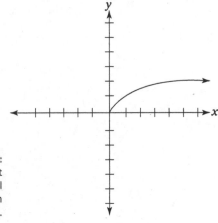

FIGURE 13-2:
The parent
radical
function
$f(x) = \sqrt{x}$.

As I discuss in Chapter 10, $f(x) = \sqrt{x}$ includes the point $(0,0)$, because $\sqrt{0} = 0$. You can think of this point as the anchor point for the radical function, which can help you to plot both vertical and horizontal transformations of this function. For example:

Which of the following is the graph of the function $f(x) = \sqrt{x+2} - 3$?

(A)

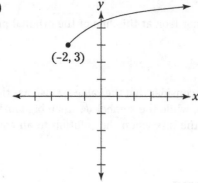

(-2, 3)

(B)

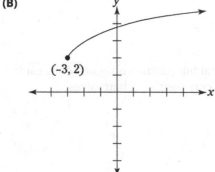

(-3, 2)

(C)

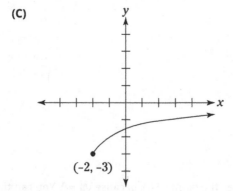

(-2, -3)

(D)

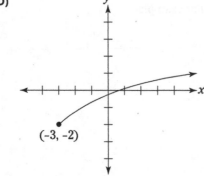

(-3, -2)

The function $f(x) = \sqrt{x+2} - 3$ is a transformation that moves the parent function $f(x) = \sqrt{x}$ two units to the left and three units down. Therefore, Answer C is correct.

Additionally, recall from Chapter 10 that the transformation $-f(x)$ reflects the function across the x-axis. Here's an SAT math question that applies this transformation to the radical function:

EXAMPLE

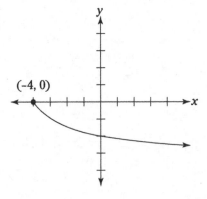

Which of the following could be the equation for this graph?

(A) $f(x) = \sqrt{-x - 4}$

(B) $f(x) = \sqrt{-x} + 4$

(C) $f(x) = -\sqrt{x + 4}$

(D) $f(x) = -\sqrt{x} - 4$

The graph shown is a transformation that moves the function $f(x) = \sqrt{x}$ four units to the left and then reflects it across the x-axis, so it's $-f(x + 4)$. Applying this transformation to $f(x) = \sqrt{x}$ results in the following:

$$-f(x + 4) = -\sqrt{x + 4}$$

Therefore, Answer C is correct.

Chapter **14**

Geometry and Trigonometry

I n this chapter, you look at two related topics: geometry and trigonometry.

To begin, I discuss the geometry that appears most often on the SAT. This includes the most common formulas for area and perimeter (circumference for a circle) and a variety of theorems for measuring angles. I also provide some basic information about triangles, especially right triangles.

This information about right triangles is the basis for trigonometry, which also appears on the SAT. Next, I show you how the trigonometric ratios sine, cosine, and tangent are constructed. I show you how to find these ratios given any angle in a right triangle. I also show you how to construct a right triangle given a single trig ratio. Most importantly, I show you how to work with an indispensible trigonometric identity, $\sin x° = \cos(90 - x)°$, which frequently appears on the SAT.

After that, you look at how degrees and radians are two contrasting ways to measure angles. I show you how to convert degrees to radians, and vice versa. Then, you use radians to calculate arc length using a formula that you may find simpler than the one you used in your math class.

Geometry

Geometry is a noble branch of mathematics dating back to the Ancient Greeks and even earlier. Euclidean geometry marks the earliest known project to ground an area of math in a set of axioms (also called postulates) from which all subsequent theorems could be rigorously constructed.

As you can see, I'm second to none in my appreciation of the triumph of geometry. And what's also true — though I may be inviting a torrent of perturbed responses in saying this — is that I personally think that ten months of high school geometry is about eight or nine months too many.

Wittingly or not, the elders of the SAT seem to agree with me. Very little of this long curriculum appears on the SAT. Setting aside trigonometry, which I deal with in the next section (and which is undeniably important for studying higher math), you can do well on the SAT by knowing a handful of geometric formulas, a few tricks for measuring angles, and the basics of working with triangles, especially right triangles.

In this section, I get you up to speed.

Geometric formulas for success

A *geometric formula* is an equation that allows you to perform a calculation using one or more known elements of a geometric shape — such as side length, height, perimeter, area, and so forth — to find an unknown element.

In some cases, an SAT math question will use a formula that's provided on the Reference list that's included on the first page of *every* SAT math section. These formulas include:

>> Area and circumference formulas for circles

>> Area formulas for rectangles and triangles

>> The Pythagorean Theorem

>> Special right triangles ($45° - 45° - 90°$ and $30° - 60° - 90°$)

>> Volume of a box, cylinder, sphere, cone, and pyramid

I discuss this Reference list further in Chapter 1. Here's an example of an SAT question that uses a formula from this list.

EXAMPLE

The radius of a standard NBA basketball is approximately 4.75 inches. To the nearest whole cubic inch, what is the volume of this basketball in cubic inches? (Use 3.14 to approximate π.)

Note that this question doesn't explicitly give you the formula for the volume of a sphere, so you have to use the formula $V = \frac{4}{3}\pi r^3$ provided in the Reference list. Plug in 4.75 for r into this formula:

$$V = \frac{4}{3}\pi r^3 = \frac{4}{3}\pi (4.75)^3$$

Now, simplify and round to the nearest whole number:

$$= \frac{4}{3}\pi (4.75)^3 = \frac{4}{3}(3.14)(107.171875) \approx 449$$

Therefore, the correct answer is 449.

In other cases, an SAT math question will provide you with the geometric formula that you'll need to answer it correctly. For example:

EXAMPLE

The formula for the surface area of a box given length *l*, width *w*, and height *h* is $A = 2(lw + lh + wh)$. If a box with a surface area of 5,380 square inches has a length of 25 inches and a width of 14 inches, what is its height in inches?

To solve this problem, plug these values into the given formula and solve for *h*:

$$5380 = 2[(25)(14) + 25h + 14h)]$$
$$5380 = 2[350 + 39h]$$
$$5380 = 700 + 78h$$
$$4680 = 78h$$
$$60 = h$$

Thus, the height of the box is 60 inches, so the correct answer is 60.

Knowing the angles

In your Geometry class, you probably learned a set of rules that allow you to find the measurements of angles. At the same time, you may have struggled to remember all the different names of these rules (alternate exterior angles, anyone?).

The good news is that on the SAT, knowing how to use a short list of these rules — and none of their obscure names — is all you need to answer most questions about angles.

Vertical angles are equal, and linear pairs add up to 180°

When two lines intersect, any pair of angles that are opposite each other are called *vertical angles*. You don't have to remember the name, but definitely remember that they're equal to each other. For example, in Figure 14-1:

» $a° = c°$

» $b° = d°$

Additionally, when two lines intersect, any pair of angles that are adjacent to each other is called a *linear pair*, and add up to 180°. For example, in Figure 14-1:

» $a° + b° = 180°$

» $b° + c° = 180°$

» $c° + d° = 180°$

» $d° + a° = 180°$

FIGURE 14-1:
Vertical angles are equal to each other; linear pairs add up to 180°.

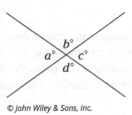

© John Wiley & Sons, Inc.

Taken together, these two rules enable you to measure all four angles around the intersection of two lines, provided that you know the measurement of one angle. For example:

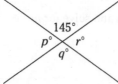

EXAMPLE

In this figure, what is the value of $q - p - r$?

To begin, angle q measures 145°. And both p and r measure 35°, so:

$$q - p - r = 145 - 35 - 35 = 75$$

Therefore, the correct answer is 75.

Corresponding angles are equal

When a line intersects two parallel lines, the two angles that are in the same position with respect to the intersecting line are called *corresponding angles*, and are equal to each other in measurement.

For example, in Figure 14-2, the following pairs are corresponding pairs and, therefore, equal:

» $a° = e°$

» $b° = f°$

» $c° = g°$

» $d° = h°$

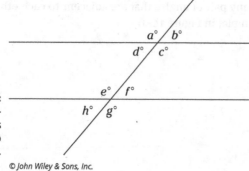

FIGURE 14-2:
Corresponding angles are equal to each other.

© John Wiley & Sons, Inc.

This single rule (when combined with the rules for vertical angles and linear pairs — see the previous section) allows you to find the measurements of all eight angles that are formed when a line intersects a pair of parallel lines, when you know the measurement of one angle. So you *really* don't need to remember the names of rules for alternate interior angles, or same-side exterior angles, to answer SAT questions. For example:

In this figure, what is the value of $u + w - v$?

To begin, angle u measures 110°. Thus, angle w also measures 110°, and angle v measures 70°, so:

$$u + w - v = 110 + 110 - 70 = 150$$

Therefore, the correct answer is 150.

The three angles in a triangle add up to 180°

This rule is pretty straightforward, and most of my SAT students remember it easily: When you add up the measures of the three angles in a triangle, the result is always 180°.

You can, of course, combine this rule with the others for angle measurement, to answer SAT questions. For example:

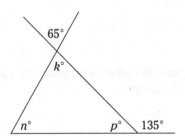

In this figure, what is the value of n?

To begin, k, n, and p are the three angles of a triangle, so you can make the following equation:

$$k + n + p = 180$$

Now, angle k and the 65° angle are vertical, so $k = 65$. Next, angle p and the 135° angle are a linear pair, so $p = 45$. So you can plug these two values into this equation as follows:

$$65 + n + 45 = 180$$
$$110 + n = 180$$
$$n = 70$$

Therefore, the correct answer is 70.

Isosceles triangles

An *isosceles triangle* has two equivalent angles and two sides of equivalent length.

Often, an SAT math question will present you with a triangle that has two equal sides, and then ask you about the angles. Remember that in an isosceles triangle, the two equal angles that match are *opposite* the equivalent sides!

An example should make this clear.

EXAMPLE

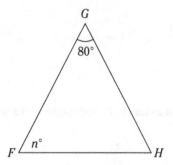

In this triangle, $FG = GH$, and angle $G = 80°$. What is the value of n?

Sides FG and GH are equal, so the triangle is isosceles. Thus, the two lower angles of the triangle are equal, so both of them measure $n°$. The three angles of a triangle add up to $180°$, so you can make and solve the following equation:

$$n + n + 80 = 180$$
$$2n + 80 = 180$$
$$2n = 100$$
$$n = 50$$

Therefore, the correct answer is 50.

When you combine this rule for isosceles triangles with the other rules you've already worked with, you can answer more complicated questions. For example:

EXAMPLE

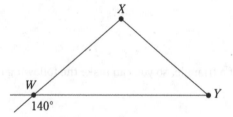

In this figure, $WX = XY$. What is the measure of angle X in degrees? (Enter your answer into the grid without the degree symbol.)

Angle YWX and the $140°$ angle are a linear pair, so $YWX = 40°$. $WX = XY$, so triangle WXY is isosceles. Thus, angle Y is also $40°$. Now, because the three angles of the triangle equal $180°$, you can make and solve the following equation:

$$W + X + Y = 180$$
$$40 + X + 40 = 180$$
$$80 + X = 180$$
$$X = 100$$

Therefore, the correct answer is $100°$, which you can enter into the grid as 100.

Similar triangles

In your Geometry class, you probably spent a lot of time working with *congruent triangles* — that is, triangles that have the same sides and angle measurements. In contrast, you probably spent a lot less time working with *similar triangles*, which have only the same angle measurements.

This is a pity, because in my humble opinion, the rules for similar triangles are vastly more useful than for congruent triangles. They're the basis for trigonometry (see later in this chapter), which itself is indispensable for land surveying, navigation, astronomy, and a ton of physics formulas that use sines and cosines to model waves.

More important for you is that SAT math questions tend to focus on similar triangles and for the most part, skip over congruent ones. The main thing you need to know is that similar triangles:

» Have three angles of equal measurement

» Have three sides whose measurements are proportional to one another

The rule of proportionality is a big deal, because it allows you to create and solve equations based on ratios (as I show you in Chapter 7). For example:

EXAMPLE

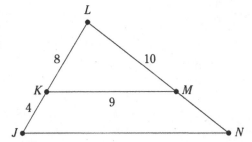

In this figure, line segment *KM* is parallel to *JN*, and the lengths of the segments are as shown. What is the length of *JN*?

The figure shows two triangles, *KLM* and *JLN*. Both triangles share angle *L*. *KM* is parallel to *JN*, so angle *LKM* and angle *J* are corresponding angles and, therefore, equivalent. Similarly, angle *LMK* and angle *N* are also equivalent. Thus, the two triangles both have the same three angles, so they're similar triangles. Therefore, their corresponding sides are proportional.

The length of *KL* is 8 and the length of *JL* is 12, so each side of triangle *JLN* is 1.5 times longer than the corresponding side of triangle *KLM*. Thus, *JN* is 1.5 times longer than *KM*, so the length of *JN* is $9 \times 1.5 = 13.5$, so the correct answer is 13.5.

In Chapter 7, when discussing ratios and proportions, I provide a few more practice problems involving similar triangles.

Working with right triangles

A *right triangle* is any triangle that includes a right angle — that is, a 90° angle. Because the three angles of every triangle add up to 180°, the remaining two angles of a right triangle also add up to 90°.

The longest side of a right triangle — the side opposite the right angle — is called the *hypotenuse*. The remaining sides are called the *legs*. In this section, I give you some key information about right triangles that you'll need for the SAT. This information is also the basis for trigonometry, which I discuss in the next section.

Right triangles and the Pythagorean Theorem

The *Pythagorean Theorem* allows you to calculate the length of any side of a right triangle when you know the lengths of the other two sides:

$$a^2 + b^2 = c^2$$

In this formula, the value c equals the length of the hypotenuse, and the values a and b can be used interchangeably for the lengths of the legs.

For example:

EXAMPLE

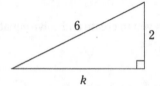

In this figure, what is the value of k?

(A) $3\sqrt{2}$

(B) $4\sqrt{2}$

(C) $3\sqrt{3}$

(D) $4\sqrt{3}$

Using the Pythagorean Theorem, substitute 6 for c and 2 for either a or b:

$$a^2 + b^2 = c^2$$
$$2^2 + b^2 = 6^2$$
$$4 + b^2 = 36$$
$$b^2 = 32$$
$$b = \sqrt{32}$$

You can factor and simplify this answer as follows:

$$\sqrt{32} = \sqrt{16}\sqrt{2} = 4\sqrt{2}$$

Therefore, Answer B is correct.

Knowing the most common Pythagorean Triples

A *Pythagorean Triple* is a set of three integers that can be used as a, b, and c values for the Pythagorean Theorem.

Although Pythagorean Triples are infinitely many, for the SAT, you can focus on the four most common ones:

>> $3 - 4 - 5$

>> $5 - 12 - 13$

» $7-24-25$

» $8-15-17$

For example, $3-4-5$ is a Pythagorean Triple because $3^2 + 4^2 = 5^2$. Thus, a right triangle can have two legs of lengths 3 and 4 and a hypotenuse of length 5.

TIP

The $3-4-5$ right triangle appears on the SAT in a wide variety of problems, and you'll almost certainly see it when you take your SAT.

EXAMPLE

On a hike, Kara started at her campsite and walked 3 kilometers due south to a lake, then turned right and walked 4 kilometers due west to the edge of a canyon. She ate lunch there, and then took a straight path back to her campsite. How many kilometers did she walk altogether?

You could solve this problem by using the Pythagorean Theorem, plugging in the values 3 and 4 for a and b. But these two values should alert you that Kara walked the perimeter of a $3-4-5$ right triangle, so she walked a total of $3 + 4 + 5 = 12$ kilometers.

Multiples of $3-4-5$ triangles also show up commonly in SAT math problems. In Table 14-1, I provide a list of the most frequently used multiples.

TABLE 14-1 The 3–4–5 Triangle and Some Multiples

Multiple	Pythagorean Triple
1	3–4–5
2	6–8–10
3	9–12–15
4	12–16–20
5	15–20–25
10	30–40–50
100	300–400–500

For example, here's a problem that's much easier to solve when you recognize multiples of the $3-4-5$ triangle:

EXAMPLE

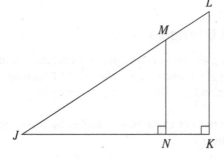

In this figure, $JK = 16$, $JL = 20$, and $MN = 9$. What is the length of ML?

Triangle *JKL* has a leg of length 16 and a hypotenuse of 20, so its remaining leg is of length 12. This triangle is similar to a 3–4–5 triangle, and so is the triangle *JNM*. *MN* has a length of 9, so this is proportional to the shortest side of a 3–4–5 right triangle. Thus, *JN* has a length of 15, so *ML* has a length of 5. Therefore, the correct answer is 5.

The next three Pythagorean Triples also show up on the SAT, so they're handy to know. In Table 14-2, I list three multiples of each of these triples.

TABLE 14-2 **The First Four Pythagorean Triples and Some Multiples**

Pythagorean Triple	Multiple of 2	Multiple of 10	Multiple of 100
5-12-13	10-24-26	50-120-130	500-1200-1300
8-15-17	16-30-32	80-150-170	800-1500-1700
7-24-25	14-48-50	70-240-250	700-2400-2500

Here's another problem that you can solve using the Pythagorean Theorem, but which is much easier to solve when you're familiar with a few common Pythagorean Triples and their multiples:

EXAMPLE

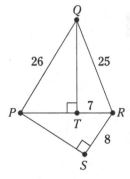

In this figure, all segments and angles are as labeled. What is the perimeter of *PQRS*?

At first glance, this problem looks like it will require a lot of calculations. But right triangle *RTQ* has a leg of length 7 and a hypotenuse of 25, so the length of *QT* is 24. So right triangle *QTP* has a leg of length 24 and a hypotenuse of 26, so the length of *PT* is 10. Thus, right triangle *RSP* has a leg of length 8 and a hypotenuse of 17, so the length of *PS* is 15. Therefore, the perimeter of *PQRS* is $26 + 25 + 8 + 15 = 74$, so the correct answer is 74.

Looking at the two special right triangles

The $45° – 45° – 90°$ and $30° – 60° – 90°$ right triangles arise so often in math that they're often called *special right triangles*. They're so special that your friends who make the SAT include them both among the Reference information that they provide on the first page of every test.

As you can see from Figure 14-3, you can construct these two triangles by bisecting a square diagonally and an equilateral triangle vertically.

The two legs and hypotenuse of a $45° – 45° – 90°$ right triangle are in a ratio of $1:1:\sqrt{2}$, as you can see in Figure 14-4. With this ratio, you can use the length of any side of this triangle to calculate the lengths of the other two sides. For example:

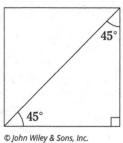

FIGURE 14-3: Constructing the two special right triangles from a square and an equilateral triangle.

© John Wiley & Sons, Inc.

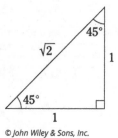

FIGURE 14-4: A $45° - 45° - 90°$ right triangle with sides of 1, 1, and $\sqrt{2}$.

© John Wiley & Sons, Inc.

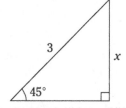

EXAMPLE

In this figure, which of the following is the best approximation of the value of *x*?

(A) 2.1

(B) 2.2

(C) 2.3

(D) 2.4

A leg and the hypotenuse of the $45° - 45° - 90°$ right triangle are in a $1 : \sqrt{2}$ ratio, so make and solve the following proportional equation:

$$\frac{x}{3} = \frac{1}{\sqrt{2}} =$$

$$\sqrt{2}x = 3$$

$$x = \frac{3}{\sqrt{2}} \approx 2.1$$

Therefore, Answer A is correct.

The two legs and hypotenuse of a $30° - 60° - 90°$ right triangle are in a $1 : \sqrt{3} : 2$ ratio, as shown in Figure 14-5. Thus, given the length of any side, you can create a proportional equation to find the lengths of the other two sides. For example:

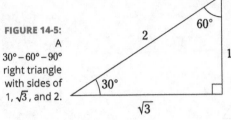

FIGURE 14-5:
A
30° − 60° − 90°
right triangle
with sides of
1, √3, and 2.

© John Wiley & Sons, Inc.

EXAMPLE

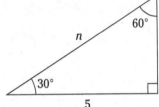

In this figure, which of the following values equals n?

(A) $5\sqrt{3}$

(B) $10\sqrt{3}$

(C) $\dfrac{5\sqrt{3}}{3}$

(D) $\dfrac{10\sqrt{3}}{3}$

The long leg and the hypotenuse of the 30° − 60° − 90° right triangle are in a $\sqrt{3} : 2$ ratio, so make and solve the following proportional equation:

$$\frac{x}{5} = \frac{2}{\sqrt{3}} =$$
$$\sqrt{3}x = 10$$
$$x = \frac{10}{\sqrt{3}}$$

This is a correct answer, but to make it look like one of the four multiple-choice answers, rationalize the denominator by multiplying the numerator and denominator by $\sqrt{3}$ and then simplifying:

$$x = \frac{10}{\sqrt{3}} \frac{\sqrt{3}}{\sqrt{3}} = \frac{10\sqrt{3}}{3}$$

Therefore, Answer D is correct.

You can combine these two ratios to answer more difficult SAT math questions. For example:

EXAMPLE

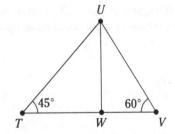

In this figure, if the length of UW is $2\sqrt{3}$, what is the length of TV?

(A) $1+\sqrt{3}$

(B) $2+\sqrt{3}$

(C) $2+2\sqrt{3}$

(D) $2+4\sqrt{3}$

Triangle UWT is a $45°-45°-90°$ right triangle, so the length of TW is $2\sqrt{3}$. Triangle VWU is a $30°-60°-90°$ right triangle, so you can find the length of WV by creating the following ratio:

$$\frac{WV}{2\sqrt{3}} = \frac{1}{\sqrt{3}} =$$
$$WV\sqrt{3} = 2\sqrt{3}$$
$$WV = 2$$

Thus, the length of TV equals $2+2\sqrt{3}$, so Answer C is correct.

Trigonometry

Trigonometry is the study of triangles, especially right triangles. It's also a major building block of higher math, and if you're headed for a college degree in the sciences, you almost can't know too much about trig. That's why most high school math curricula spend a lot of time on this topic.

That said, if your current math class is Algebra II, Pre-Calculus, or beyond, you may well have already studied all the trigonometry you need to know for the SAT. In this section, I identify and review this subset of information to make sure you haven't misplaced it along the way.

Using SOH-CAH-TOA to understand sines, cosines, and tangents

You know from earlier in this chapter that every right triangle has a right angle that measures 90°. The longest side of a right triangle is called the hypotenuse, which is identified in trigonometry by the capital letter H. The remaining two sides are called legs.

The study of trigonometry begins by identifying a *reference angle* θ in a right triangle — that is, an angle that's used to distinguish the two legs of the triangle. The leg opposite θ is called the *opposite* side of the triangle, O; the remaining leg is called the *adjacent* side, A. Figure 14-6 shows the positions of these three sides of a right triangle in relation to the reference angle.

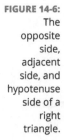
FIGURE 14-6:
The opposite side, adjacent side, and hypotenuse side of a right triangle.

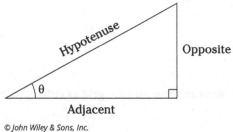
© John Wiley & Sons, Inc.

With these distinctions in place, the key tools that trigonometry uses to study triangles are the *trigonometric ratios*: the *sine*, *cosine*, and *tangent*, abbreviated *sin*, *cos*, and *tan*, respectively:

$$\sin\theta = \frac{O}{H} \qquad\qquad \cos\theta = \frac{A}{H} \qquad\qquad \tan\theta = \frac{O}{A}$$

This information is summed up as the mnemonic SOH-CAH-TOA.

You can find the sine, cosine, or tangent of any angle θ by calculating the correct ratio of two sides of a right triangle that includes that angle.

For example, consider the trusty 3–4–5 right triangle in Figure 14-7.

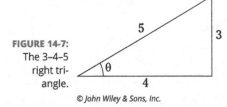

When the reference angle θ is the angle between the hypotenuse and the leg that measures 4, the opposite side O measures 3 and the adjacent side A measures 4. Thus, you can calculate the sine, cosine, and tangent as follows:

$$\sin\theta = \frac{O}{H} = \frac{3}{5} = 0.6 \qquad \cos\theta = \frac{A}{H} = \frac{4}{5} = 0.8 \qquad \tan\theta = \frac{O}{A} = \frac{3}{4} = 0.75$$

For example:

EXAMPLE

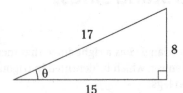

In this triangle, what is the value of $\cos\theta - \sin\theta$?

In this triangle, the opposite side measures 8, the adjacent side measures 15, and the hypotenuse measures 17. Use these three values to calculate the cosine and sine of θ:

$$\cos\theta = \frac{A}{H} = \frac{15}{17} \qquad\qquad\qquad \sin\theta = \frac{O}{H} = \frac{8}{17}$$

Now, calculate $\cos\theta - \sin\theta$:

$$\cos\theta - \sin\theta = \frac{15}{17} - \frac{8}{17} = \frac{7}{17}$$

Therefore, the correct answer is $\frac{7}{17}$, which you can enter into the grid as either .411 or .412.

Building a triangle from a single trig ratio

When you know the value of one trig ratio for a given angle θ, you can calculate all the remaining trig ratios for θ. For example:

EXAMPLE

If θ is an acute angle and $\tan\theta = \frac{7}{24}$, what is $\sin\theta$?

(A) $\frac{7}{24}$

(B) $\frac{7}{25}$

(C) $\frac{24}{25}$

(D) $\frac{25}{24}$

The angle θ is an acute angle, so you can draw a right triangle that includes θ as a reference angle.

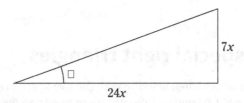

The value $\tan\theta = \frac{7}{24}$, so the opposite and adjacent sides of this triangle are in a 7:24 ratio, and so I label these sides $7x$ and $24x$. (Don't let the x throw you! It's essentially a placeholder for some multiple that changes the size of the triangle without affecting the angles or the trig ratios.) You may recognize this triangle as the Pythagorean Triple 7–24–25, but if not, you can use the Pythagorean Theorem to find the length of the hypotenuse:

$$a^2 + b^2 = c^2$$
$$7^2 + 24^2 = c^2$$
$$49 + 576 = c^2$$
$$625 = c^2$$
$$25 = c$$

Thus, the length of the hypotenuse is $25x$. You can use this value to find $\sin\theta$:

$$\sin\theta = \frac{O}{H} = \frac{7x}{25x} = \frac{7}{25}$$

As you can see, the x-factor cancels out of the ratio, so Answer B is correct.

In some cases, the problem will present you with a triangle that's a multiple of one of the Pythagorean Triples that I cover earlier in this chapter. For example:

EXAMPLE

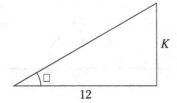

In this triangle, if θ is an acute angle and $\cos\theta = \frac{4}{5}$, then solve for k.

The trig ratio $\cos\theta = \dfrac{A}{H}$, and the figure shows that $A = 12$, so you can use the following equation to find the hypotenuse:

$$\frac{12}{H} = \frac{4}{5}$$
$$60 = 4H$$
$$15 = H$$

You may now recognize this triangle as the 9−12−15 triangle, which is a multiple of the 3−4−5 triangle. If not, use the Pythagorean Theorem to solve for k:

$$a^2 + b^2 = c^2$$
$$12^2 + k^2 = 15^2$$
$$144 + k^2 = 225$$
$$k^2 = 81$$
$$k = 9$$

Therefore, the correct answer is 9.

Applying trig ratios to special right triangles

An important application of the trig ratios arises when working with the two special right triangles, the $45° - 45° - 90°$ and $30° - 60° - 90°$ right triangles. I discuss these two triangles earlier in this chapter. For convenience, they're reproduced here in Figure 14-8.

FIGURE 14-8: The $45° - 45° - 90°$ and $30° - 60° - 90°$ right triangles.

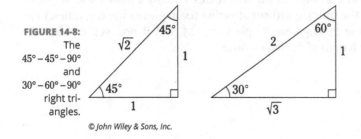

© John Wiley & Sons, Inc.

Table 14-3 shows the sine, cosine, and tangent when θ equals 30°, 45°, and 60°.

TABLE 14-3 Sine, cosine, and tangent of three common angles

30°	45°	60°
$\sin 30° = \dfrac{1}{2}$	$\sin 45° = \dfrac{1}{\sqrt{2}} = \dfrac{\sqrt{2}}{2}$	$\sin 60° = \dfrac{\sqrt{3}}{2}$
$\cos 30° = \dfrac{\sqrt{3}}{2}$	$\cos 45° = \dfrac{1}{\sqrt{2}} = \dfrac{\sqrt{2}}{2}$	$\cos 60° = \dfrac{1}{2}$
$\tan 30° = \dfrac{1}{\sqrt{3}} = \dfrac{\sqrt{3}}{3}$	$\tan 45° = 1$	$\tan 60° = \sqrt{3}$

EXAMPLE

These values arise so frequently in trigonometry that you may well face one or more SAT math questions that include them. For example:

If $\cos\theta = \frac{1}{2}$ and $0° \leq \theta \leq 90°$, then $\tan(\theta - 15°) =$

(A) 1

(B) $\sqrt{3}$

(C) $\frac{\sqrt{3}}{2}$

(D) $\frac{\sqrt{3}}{3}$

You know that $\cos 60° = \frac{1}{2}$, so $\theta = 60°$, and thus $\theta - 15° = 45°$. Thus, because $\tan 45° = 1$, Answer A is correct.

Remembering the most important SAT identity

If you remember only one thing about trigonometry for your SAT, make it this identity:

$$\sin x° = \cos(90 - x)°$$

To understand why this identity is true, look at the triangle in Figure 14-9.

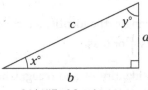
As you can see, the two angles of this triangle measure $x°$ and $y°$, with the lengths of the two legs a and b, and the hypotenuse c. Now, here are the four values for the sine and cosine of x and y:

$$\sin x° = \frac{a}{c} \qquad\qquad \sin y° = \frac{b}{c}$$

$$\cos x° = \frac{b}{c} \qquad\qquad \cos y° = \frac{a}{c}$$

Notice that regardless of the values of a, b, and c, these two equations emerge:

$$\sin x° = \cos y° \qquad\qquad \sin y° = \cos x°$$

Now, to complete the picture, recall from geometry that the three angles of a triangle always add up to 180°. And because a right triangle always includes one angle that measures 90°, the two smaller angles of a right triangle always add up to 90°. Thus:

$$x° + y° = 90°$$

$$y° = 90° - x°$$

So, you can replace $y°$ with $90° - x°$ in both of these equations. Therefore, you can rewrite them as:

$$\sin x° = \cos(90° - x°) \qquad\qquad \sin(90° - x°) = \cos x°$$

Both of these results include essentially the same information, and are encapsulated in the first version of it. I've taken all these pains to show you this stuff because you'll almost certainly be asked one question on your SAT that can be answered quickly and easily if you know this identity. Here are a few examples to hammer the point home:

EXAMPLE

If $\sin\theta° = n$, which of the following is also true for all values of x?

(A) $\cos x° = -n$

(B) $\cos(\theta + a)° = n + a$

(C) $\cos(90 - \theta)° = n$

(D) $\cos(90 - \theta)° = -n$

The identity $\sin x° = \cos(90° - x°)$ leads directly to the conclusion that $\cos(90 - \theta)° = n$, so Answer C is correct.

EXAMPLE

$$\sin 20° = \cos k°$$

If this equation is true and $0 < k < 90$, what is the value of k?

The quick answer is that $\cos k°$ can only equal $\sin 20°$ if the sum of the two angles is 90°, so $k = 70$.

EXAMPLE

If the two acute angles in a right triangle measure $u°$ and $2v°$, and $\sin u° = \frac{1}{4}$, what is the value of $\cos 2v°$?

Again, because $u°$ and $2v°$ are the two angles in a right triangle, the identity $\sin u° = \cos 2v°$ holds in this case, so $\cos 2v° = \frac{1}{4}$; therefore, the correct answer is $\frac{1}{4}$ or 0.25.

I hope these three examples illustrate how knowing this identity — and recognizing when and how to use it! — can get you a quick point on the SAT almost for free.

Radian measure and arc length

Trigonometry uses *radian measure* — angle measurement based on the circumference of a unit circle — to simplify difficult calculations. Radian measure can be confusing to work with at first. Fortunately, you can answer a lot of SAT questions with only a basic knowledge of how to convert between degrees and radians.

After you understand how to perform basic conversions, you're ready to face down questions about arc length. To help you get through this, I give you a version of the arc length formula that is simpler to remember than the one you probably remember from trig class.

Understanding radian measure for angles

Radian measure arises naturally from measurements with angles and circles. To see how this happens, consider the unit circle — that is, a circle centered on the xy-graph with a radius of 1, as shown in Figure 14-10.

Recall that the number of degrees in the arc of a circle is 360°. Additionally, the circumference of a circle equals $2\pi r$. Thus, a unit circle has a radius of 1, so it has a circumference of 2π. Radian measure begins with this fundamental equation of the number of degrees and the arc length of a unit circle:

$$360° = 2\pi$$

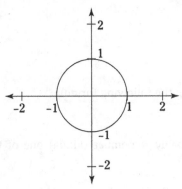

FIGURE 14-10:
The unit circle.

© John Wiley & Sons, Inc.

Dividing both sides of this equation by 2 simplifies this equation still further:

$$180° = \pi$$

Further divisions provide equations for four other important angles:

$$90° = \frac{\pi}{2} \qquad\qquad 60° = \frac{\pi}{3}$$

$$45° = \frac{\pi}{4} \qquad\qquad 30° = \frac{\pi}{6}$$

You can use the variations of the form $180° = \pi$ to create the two conversion factors that allow you to convert degrees to radians and vice versa:

To convert degrees to radians: $\frac{\pi}{180°}$ To convert radians to degrees: $\frac{180°}{\pi}$

Here's an example of how to use the first conversion factor to convert degrees to radians:

EXAMPLE

What is the equivalent of 35° in radian measure?

(A) $\frac{7\pi}{24}$

(B) $\frac{7\pi}{30}$

(C) $\frac{7\pi}{36}$

(D) $\frac{7\pi}{48}$

To convert 35° to radians, multiply by the conversion factor $\frac{\pi}{180°}$:

$$35° \times \frac{\pi}{180°} = \frac{35\pi}{180} = \frac{7\pi}{36}$$

Therefore, Answer C is correct.

As another example, here's how you convert radians to degrees:

EXAMPLE

If $\frac{3\pi}{5}$ radians is equivalent to $x°$, then what value does x equal?

This time, multiply by the conversion factor $\frac{180°}{\pi}$:

$$\frac{3\pi}{5} \times \frac{180°}{\pi} = \frac{540°}{5} = 108°$$

Therefore, the correct answer is 108 degrees.

Measuring arc length using radians

To find the arc length of a circle, I recommend using the following formula:

Arc length = radius × radians

If this formula looks a little unfamiliar, you probably remember (kinda) one of these more unwieldy versions from your trig class:

$$\text{Arc length} = 2\pi r \frac{\theta}{360} \qquad\qquad \text{Arc length} = \pi r \frac{\theta}{180}$$

As an SAT teacher, I can tell you that I've rarely met a student who could recite one of these two formulas confidently from memory. When I tell them there's an easier alternative, they're usually happy to hear about it.

The one catch here is that you need to remember how to convert degrees to radians (which I show you in the previous section). But most of my students are actually pretty good about remembering the conversion factor $\frac{\pi}{180°}$. Here's an example to show you how to use the simplified arc length formula:

EXAMPLE

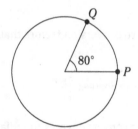

If this circle has a radius of 6, what is the arc length from P to Q?

(A) $\frac{8\pi}{3}$

(B) $\frac{9\pi}{4}$

(C) $\frac{11\pi}{5}$

(D) $\frac{13\pi}{6}$

To begin, convert 80° to radians:

$$80° \times \frac{\pi}{180°} = \frac{80\pi}{180} = \frac{4\pi}{9}$$

Next, plug in 6 for the radius and $\frac{4\pi}{9}$ for radians into the formula for arc length:

$$\text{Arc length} = 6 \times \frac{4\pi}{9} = \frac{24\pi}{9} = \frac{8\pi}{3}$$

Therefore, Answer A is correct.

Chapter **15**

Additional SAT Math Topics

This chapter rounds out the most advanced math that you'll need in order to do well on the SAT math test, with a grab-bag of topics that don't fit neatly into other chapters in this part of the book.

First, I show how imaginary numbers arise from the observation that $i = \sqrt{-1}$ and its multiples don't fit neatly onto the real number line along with other numbers you've worked with. You get comfortable performing a variety of calculations using i, including raising it to the power of a real integer.

Then, you work with complex numbers, which are numbers of the form $a + bi$. You discover how to answer SAT math questions that require you to add, subtract, multiply, and divide complex numbers.

In the final section, you work with circles on the *xy*-plane, using the two formulas $x^2 + y^2 = r^2$ and $(x-h)^2 + (y-k)^2 = r^2$ to answer some of the most common types of SAT math questions within this topic. I also show you how to complete the square to answer difficult questions involving graphs of circles.

Imaginary and Complex Numbers

Most SATs include a question or two about imaginary numbers (multiples of i, which equals $\sqrt{-1}$) and the set of complex numbers (numbers of the form $a + bi$). In this section, I start by giving you a grounding in how i is defined and used in some basic calculations. Then, I introduce you to the complex numbers, and show you how to answer the most common types of SAT math questions that include them.

The imaginary number *i*

The imaginary number *i* seems mysterious and hard to understand, so some students shy away from answering SAT math questions that include this value. In this section, I hope to demystify *i* to help you understand how to work with it, especially in the context of complex numbers.

Imagining and defining *i*

An *imaginary number* is a number that is a multiple of $\sqrt{-1}$, which is represented by the letter *i*. Thus, by definition:

$$i = \sqrt{-1}$$

An equally valid way of defining this value is:

$$i^2 = -1$$

This value, as well as all imaginary numbers, is *not* a real number — that is, it's not on the real number line. Considering that other strange numbers — for example, $\sqrt{2}$, π, and *e* — exist happily with other values on the real number line, why is $\sqrt{-1}$ so different?

To understand the strangeness of $\sqrt{-1}$, remember that taking the square root of a number is the inverse of squaring a number, which is simply multiplying a number by itself. So, to find $\sqrt{-1}$, you need to find a number *n* that satisfies the following equation:

$$n \times n = -1$$

But if this value *n* is on the real number line, shown in Figure 15-1, then it must be a positive number, a negative number, or 0. However:

>> The value *n* isn't positive, because when you multiply a positive number by itself, the result is positive.

>> The value *n* isn't negative, because when you multiply a negative number by itself, the result is positive.

>> The value *n* isn't 0, because $0 \times 0 = 0$.

FIGURE 15-1: The real number line.

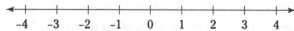

The unmistakable conclusion is that if $\sqrt{-1}$ exists as a number at all, then this number isn't on the real number line. For centuries, mathematicians avoided $\sqrt{-1}$ and its multiples. In the 17th century, mathematician and philosopher René Descartes, the inventor of the *xy*-plane, called these numbers *imaginary* to disparage their existence. In the 18th century, mathematician Leonhard Euler adopted the notation *i* to stand for this value, and his breakthrough work with it convinced his peers that imaginary numbers were an indispensable mathematical tool.

Calculating with *i*

When you accept *i* as a number, you're ready to do some basic calculations with this unusual value. In this section, I get you up to speed on how to add, subtract, and multiply with imaginary numbers — that is, any term that includes *i*. I also show you how to raise *i* to an integer exponent.

ADDING AND SUBTRACTING WITH *I*

When you add *i* to itself, the result is as you might expect:

$$i+i=2i \qquad\qquad i+i+i=3i \qquad\qquad i+i+i+i=4i$$

Thus, you can combine imaginary like terms using the same algebra rules that you use with real numbers:

$$3i+4i=7i \qquad\qquad 10i+i+i=12i \qquad\qquad 3000i+300i+30i+3i=3333i$$

Additionally (or maybe subtractionally?), you can also subtract imaginary like terms:

$$8i-5i=3i \qquad\qquad 20i-i-i=18i \qquad\qquad 100i-22i=78i$$

Subtraction can also lead to negative imaginary values:

$$5i-8i=-3i \qquad\qquad 2i-i-i=0 \qquad\qquad 100i-1000i=-900i$$

MULTIPLYING WITH I

From the previous section, you can see that when you multiply *i* by a real number, the notation is essentially the same as in real number algebra:

$$i\times 6=6i \qquad\qquad 2i\times 5\times 3=30i \qquad\qquad 19i\times 4=78i$$

You can also multiply positive and negative imaginary numbers by positive and negative imaginary real numbers using the rules that you already know from real number algebra:

$$-i\times 4=-4i \qquad\qquad 6i\times(-7)=-42i \qquad\qquad -5i\times-12=60i$$

Multiplying $i\times i$ may seem strange at first, but recall that $i^2=-1$ according to a basic definition of *i*. Therefore:

$$i\times i=i^2=-1$$

RAISING I TO AN EXPONENT

By definition, you know these two facts about exponents of *i*:

$$i=\sqrt{-1}$$
$$i^2=-1$$

With these facts, you can extrapolate other integer exponents of *i*. For example:

$$i^3=i\times i\times i=i\times(-1)=-i$$
$$i^4=i\times i\times i\times i=(-1)(-1)=1$$

Because $i^4=1$, subsequent exponents of *i* follow this same pattern:

$$i^5=i \qquad\quad i^6=-1 \qquad\quad i^7=-i \qquad\quad i^8=1$$
$$i^9=i \qquad\quad i^{10}=-1 \qquad\quad i^{11}=-i \qquad\quad i^{12}=1$$

This pattern continues indefinitely. Note that every exponent of *i* that's divisible by 4 equals 1, so:

$$i^{16}=1 \qquad\qquad i^{20}=1 \qquad\qquad i^{24}=1 \qquad\qquad i^{28}=1$$

This pattern enables you simplify many difficult exponents of i by subtracting a multiple of 4 from the exponent. For example:

EXAMPLE

What is the simplified value of i^{99}?

(A) 1

(B) −1

(C) i

(D) −i

You can subtract any multiple of 4 from the exponent and then simplify:

$$i^{99} = i^{99-96} = i^3 = -i$$

Thus, Answer D is correct.

This pattern also carries into negative and zero exponents of imaginary numbers, so:

$$i^{-3} = i \qquad i^{-2} = -1 \qquad i^{-1} = -i \qquad i^0 = 1$$

TIP

Note that $i^{-1} = -i$ and *not i*, as you may expect. To keep this clear, a good way to evaluate a negative exponent of i is to *add* a multiple of 4 to the exponent. For example:

EXAMPLE

What is the simplified value of i^{-27}?

(A) 1

(B) −1

(C) i

(D) −i

To solve, add a helpful multiple of 4 to the exponent and then simplify:

$$i^{-27} = i^{-27+28} = i^1 = i$$

Thus, Answer C is correct.

Complex numbers

In the previous section, you apply a variety of mathematical operations to imaginary numbers. When you add (or subtract) a real number and an imaginary number, however, the result is a *complex number* that cannot be further simplified. Thus, every complex number $(a + bi)$ has a real part a and an imaginary part bi.

Questions about operations on complex numbers show up regularly on the SAT. In this section, I show you how to add, subtract, multiply, and divide pairs of complex numbers.

Adding and subtracting complex numbers

Adding and subtracting complex numbers is one of the simplest things you may be asked to do on the SAT. To add a pair of complex numbers, simply combine like terms, adding the real parts of each number and then separately adding the imaginary parts. For example:

EXAMPLE

What is the sum of $3+7i$ and $4-9i$?

(A) $5i$

(B) $-5i$

(C) $7+2i$

(D) $7-2i$

To add $3+7i$ and $4-9i$, make a single expression and combine the two real parts and the two imaginary parts just as you would combine like terms in algebra with real numbers:

$$3+7i+4-9i=7-2i$$

And so, Answer D is correct.

Subtracting a pair of complex numbers is almost as simple — just be sure to use parentheses to make sure you subtract the *entire* second number. For example:

EXAMPLE

When you subtract $8-3i$ minus $-2+9i$, what is the result?

(A) $6+6i$

(B) $6-12i$

(C) $10+6i$

(D) $10-12i$

To subtract $8-3i$ minus $-2+9i$, make a single expression using parentheses for the second complex number:

$$8-3i-(-2+9i)$$

Now, distribute the minus sign and remove the parentheses:

$$=8-3i+2-9i$$

To finish, combine like terms — that is, the two real parts and the two imaginary parts:

$$=10-12i$$

So Answer D is correct.

Multiplying complex numbers

Multiplying a pair of complex numbers is a lot like FOILing, which I discuss in Chapter 3. For example:

EXAMPLE

What is $4+3i$ multiplied by $2-5i$?

(A) $-7+14i$

(B) $-7-14i$

(C) $23+14i$

(D) $23-14i$

To multiply $(4+3i)(2-5i)$, multiply each term by every other term, as you would when FOILing a pair of binomials:

$$(4+3i)(2-5i)=8-20i+6i-15i^2$$

Next, combine the two i terms:

$$= 8 - 14i - 15i^2$$

Although this looks complete, you're not done. Recall that $i^2 = -1$, so you can substitute -1 for i^2 into the final term and simplify:

$$= 8 - 14i - 15(-1)$$
$$= 8 - 14i + 15$$
$$= 23 - 14i$$

Therefore, Answer D is correct.

Dividing complex numbers

Dividing is probably the most difficult complex number operation you'll have to do on the SAT. To divide a pair of complex numbers, you need to know how to multiply them, as I show you in the preceding section. You also need to know how to find the complex conjugate of a complex number, which is super easy. If you can do those two things, you can probably knock off any SAT math question involving complex number division.

FINDING THE COMPLEX CONJUGATE

Every complex number has a *complex conjugate*, which is simply that number with the sign of its imaginary part flipped — that is, changed either from plus to minus or from minus to plus.

Another way to say this is that complex numbers come in pairs called *conjugate pairs*. (I also discuss conjugate pairs in Chapter 12, when discussing complex roots of quadratic equations.)

When you multiply a pair of complex conjugates, the result is always a real number. Even better, this value is easy to calculate: Just square the coefficients of each part of the complex number, and add these values together.

To see why this works, I'll use the following example to walk you through:

EXAMPLE

When the complex number $5 - 4i$ is multiplied by its complex conjugate, what is the result?

(A) 20

(B) 20i

(C) 41

(D) 41i

The conjugate of $5 - 4i$ is $5 + 4i$. You can multiply these values by FOILing and then simplify:

$$(5 - 4i)(5 + 4i) = 25 + 20i - 20i - 16i^2 = 25 - 16i^2$$

As you can see, the two i-terms cancel each other. This always happens when multiplying conjugate pairs. Now, just substitute -1 for i^2, as I outline earlier in this section, and simplify:

$$= 25 - 16(-1) = 25 + 16 = 41$$

Therefore, Answer C is correct. Notice that you could have saved a lot of time and effort on this problem by simply pulling out the coefficients 5 and 4, squaring them both, and adding the result:

$$5^2 + 4^2 = 25 + 16 = 41$$

Table 15-1 shows a variety of examples where complex conjugate pairs are multiplied. Knowing how to multiply complex conjugate pairs quickly is useful when dividing complex numbers, as I cover in the next section.

TABLE 15-1 Multiplying a Complex Number and Its Conjugate
Always Results in a Real Number

Complex Number	Complex Conjugate	Multiplying Complex Conjugate Pairs
$2+3i$	$2-3i$	$(2+3i)(2-3i)=4+9=13$
$1-5i$	$1+5i$	$(1-5i)(1+5i)=1+25=26$
$-4+8i$	$-4-8i$	$(-4+8i)(-4-8i)=16+64=80$
$-6-i$	$-6+i$	$(-6-i)(-6+i)=36+1=37$

USING THE COMPLEX CONJUGATE TO DIVIDE

The complex conjugate is indispensable for dividing a pair of complex numbers. Here's an example that shows you the procedure for dividing complex numbers:

EXAMPLE

The expression $\dfrac{1-5i}{3-2i}$ is equivalent to which of the following?

(A) $1+i$

(B) $1-i$

(C) $-1+i$

(D) $-1-i$

To solve, multiply both the numerator and denominator of $\dfrac{1-5i}{3-2i}$ by the complex conjugate of the denominator, which is $3+2i$.

$$\frac{1-5i}{3-2i}=\frac{(1-5i)(3+2i)}{(3-2i)(3+2i)}$$

Evaluate the numerator by FOILing and simplifying:

$$(1-5i)(3+2i)=3+2i-15i-10i^2=3-13i-10(-1)=13-13i$$

You can evaluate the denominator even more easily using the trick I show you in the previous section for multiplying conjugate pairs:

$$(3-2i)(3+2i)=3^2+2^2=9+4=13$$

Now, substitute these two values into the numerator and denominator of the expression you're evaluating, and simplify:

$$\frac{(1-5i)(3+2i)}{(3-2i)(3+2i)}=\frac{13-13i}{13}=1-i$$

Thus, Answer B is correct.

Circles on the *xy*-Plane

On just about every SAT math test, at least one question requires you to know something about graphing circles on the *xy*-plane. If you'd like to have a good shot at answering this question correctly when you take the SAT, read on!

In this section, I show you two basic versions of center-radius form for circles, and how to work with these two forms to answer the most common types of SAT math questions. I also discuss how to answer tougher questions by changing these equations to inequalities. And I show you how to change circles whose equations are in standard form to the more useful center-radius form.

Circles in center-radius form

The most useful form that the equation of a circle can take is called *center-radius form*. In this section, I show you two related versions of this form. First, I show you the simplified version, for circles centered at the origin (0,0). After that, I show you the more complex version that allows you to plot any circle on the *xy*-plane.

Circles centered at the origin

The center-radius form for a circle centered at the origin with a radius of length *r* is:

$$x^2 + y^2 = r^2$$

TIP

Notice that this equation looks a bit like the Pythagorean Theorem: The sum of two squares equals a third square. If you like, you can use this similarity as a mnemonic to help you remember the form of this equation.

In practice, the *r* value is usually filled in as a constant. For example, Figure 15-2 shows two examples of circles on the graph, the first with a radius of 3 and the second with a radius of 6.

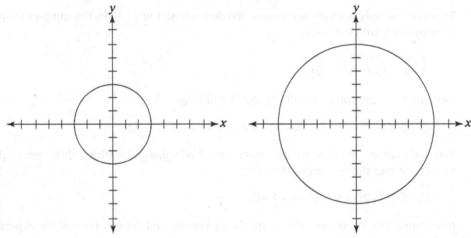

FIGURE 15-2:
Two
examples
of circles on
the *xy*-plane.

For example:

EXAMPLE

If $x^2 + y^2 = 4$, what is the radius of the resulting circle when graphed on the *xy*-plane?

The value 4 in the equation equals r^2, where *r* is the radius, so

$$r^2 = 4$$
$$r = 2$$

Therefore, the correct answer is 2.

TIP

When the radius of a circle is a whole number, the value of r^2 is a square number. Similarly, when the radius of a circle is a fraction, the value of r^2 is a fraction whose numerator and denominator are both square numbers.

These values often — not *always*, but *often!* — can tip in the direction of a correct answer.

EXAMPLE

For example:

Which of the following is the equation for a circle centered at the origin with a radius of $\frac{7}{5}$?

(A) $x^2 + y^2 = \frac{7}{5}$

(B) $x^2 - y^2 = \frac{7}{5}$

(C) $x^2 + y^2 = \frac{49}{25}$

(D) $x^2 - y^2 = \frac{49}{25}$

If $r = \frac{7}{5}$, then $r^2 = \frac{49}{25}$. Note that both 49 and 25 are square numbers, so Answers C and D are definitely looking good. Recalling that the basic formula for the graph of a circle resembles the Pythagorean Theorem should help you rule out Answer D, so Answer C is correct.

Circles centered at any point on the xy-plane

The center-radius form of the equation for a circle with a radius of length r and centered at the point (h,k) is:

$$(x-h)^2 + (y-k)^2 = r^2$$

WARNING

Be careful when plugging values into this equation! For example, suppose you want to write the equation for a circle with a radius of 4 centered at $(-1,3)$ into the equation. When you plug the values $h = -1$, $k = 3$, and $r = 4$ into the equation, here's the result:

$$(x+1)^2 + (y-3)^2 = 16$$

As you can see, the signs of both -1 and 3 are negated in the formula. SAT math questions often offer you a variety of wrong answer choices based on this fine point! For example:

EXAMPLE

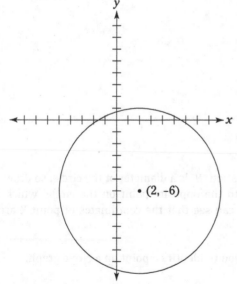

This figure shows a circle with a radius of 7 centered at $(2,-6)$. Which of the following is the equation of this graph?

(A) $(x+2)^2 + (y+6)^2 = 49$

(B) $(x+2)^2 + (y-6)^2 = 49$

(C) $(x-2)^2 + (y+6)^2 = 49$

(D) $(x-2)^2 + (y-6)^2 = 49$

Plug in 2 for h, -6 for k, and 7 for r into the formula $(x-h)^2 + (y-k)^2 = r^2$, and simplify:

$$(x-2)^2 + (y-(-6))^2 = 7^2$$
$$(x-2)^2 + (y+6)^2 = 49$$

Therefore, Answer C is correct.

Plotting points on a circle on the *xy*-plane

An SAT question about circles on the *xy*-plane might ask you not just to identify the center point or radius of that circle, but also to find points on the circle itself. For example:

$$(x+2)^2 + (y-3)^2 = 9$$

In the *xy*-plane, the graph of this equation is a circle. Point J on this circle has coordinates $(1,3)$. If JK is a diameter of the circle, then what are the coordinates of point K?

(A) $(7,3)$

(B) $(-2,0)$

(C) $(-2,6)$

(D) $(-5,3)$

From your knowledge of circles, you know that this circle is centered at $(-2,3)$ and has a radius of 3. Point J is at $(1,3)$. From this information, you can make the following sketch:

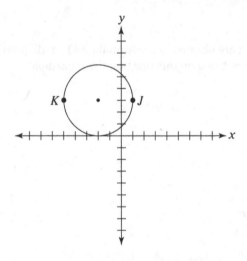

Now you're ready to take a stab at the question: JK is a diameter of the circle, so draw a line from point J through the center of the circle to the opposite point on the circle, which is point K. Because the radius of the circle is 3, you can see that the coordinates of point K are $(-5,3)$, so Answer D is correct.

Here's a final example that also requires you to identify a point on a circle graph:

$$(x-4)^2 + (y+3)^2 = 25$$

In the *xy*-plane, the graph of the equation is a circle. If OP is a diameter of the circle and point O is the origin, then what are the coordinates of point P?

(A) $(5,1)$

(B) $(5,-8)$

(C) $(8,-6)$

(D) $(9,-3)$

This time, the circle is centered at (4,–3) and has a radius of 5. Point O is the origin (0,0). From this information, you can make the following sketch:

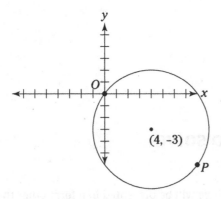

OP is a diameter of the circle, so draw a line from point O through the center of the circle to the opposite point on the circle, which is point P. Because the radius of the circle is 5, you can see that the coordinates of point P are (8,–6), so Answer C is correct.

Identifying interior and exterior points on a circle

When an SAT question asks you to identify either an interior or an exterior point of a circle graphed on the xy-plane, you can try sketching the circle using the methods I provide earlier in this section.

Alternatively, you can modify the center-radius form of the equation, changing it to an inequality as follows.

Interior points: $(x-h)^2 + (y-k)^2 < r^2$

Exterior points: $(x-h)^2 + (y-k)^2 > r^2$

For example:

$$(x+1)^2 + (y-2)^2 = 16$$

When this equation is graphed as a circle on the xy-plane, which of the following points is exterior to that circle?

(A) (1,5)

(B) (–3,3)

(C) (–4,–1)

(D) (1,–1)

To answer the question without sketching the graph, change the equals sign in the equation to a greater-than sign (>):

$$(x+1)^2 + (y-2)^2 > 16$$

Now, plug each of the four (x,y) pairs into this inequality. Only $(-4,-1)$ satisfies the inequality, as follows:

$$(-4+1)^2 + (-1-2)^2 > 16$$
$$(-3)^2 + (-3)^2 > 16$$
$$9 + 9 > 16$$
$$18 > 16$$

Thus, Answer C is correct.

Completing the square to solve difficult circle problems

EXAMPLE

In some cases on the SAT, an equation for a circle will be presented in a form other than center-radius form. For example:

$$x^2 + 6x + y^2 - 4y = 36$$

When plotted on the xy-plane, the graph of this equation is a circle with a radius of r. What is the value of r?

Because the equation isn't presented in the proper form, you can't assume that the right side of the equals sign equals r^2, so the radius *isn't* 6, as you may have thought.

In this case, your best bet is to put the equation into the center-radius form by *completing the square* for both the x and y variables. In Chapter 12, I show you how to complete the square to change a quadratic function from standard form to vertex form. Here, you use essentially the same procedure to convert the equation of a circle.

To convert this equation to center-radius form, begin by completing the square for the x-variable. To do this, take the coefficient 6, divide it by 2, and then square the result:

$$\left(\frac{6}{2}\right)^2 = 3^2 = 9$$

Add 9 to both sides of the equation:

$$(x^2 + 6x + 9) + y^2 - 4y = 36 + 9$$

I also use parentheses to show how the first three terms of the left side are meant to be grouped. This step isn't strictly necessary, but it may help you to see how the equation is evolving. Now, do the same for the coefficient of y, which is -4:

$$\left(\frac{-4}{2}\right)^2 = (-2)^2 = 4$$

So, add another 4 to both sides of the equation:

$$(x^2 + 6x + 9) + (y^2 - 4y + 4) = 36 + 9 + 4$$

Again, I use parentheses to group the last three terms on the left side. Now, for the final step, factor the contents of both sets of parentheses and add the three values on the right side:

$$(x+3)^2 + (y+2)^2 = 49$$

As a result, the equation is converted to center-radius form. Now, you can see that $r^2 = 49$, so the radius of the circle is 7.

5

Practice Makes Perfect: SAT Math Section Practice Tests

Chapter **16**

Practice SAT Math Test 1

Ready to put your preparation to the test? This practice test includes two math sections, just like the ones you'll face on your actual SAT. Work through each section in the time stated at the top of that section.

As on the SAT, you may **not** use a calculator for the first math section, but you may use one for the second. I also recommend that you tear out and use the answer sheet provided on the next page.

When you've completed both sections, turn to Chapter 17 for answers and complete explanations.

Good luck!

Answer Sheet for Practice SAT Math Test 1

Use the ovals and grid-ins provided with this practice exam to record your answers.

Section 3

1. Ⓐ Ⓑ Ⓒ Ⓓ   4. Ⓐ Ⓑ Ⓒ Ⓓ   7. Ⓐ Ⓑ Ⓒ Ⓓ   10. Ⓐ Ⓑ Ⓒ Ⓓ   13. Ⓐ Ⓑ Ⓒ Ⓓ
2. Ⓐ Ⓑ Ⓒ Ⓓ   5. Ⓐ Ⓑ Ⓒ Ⓓ   8. Ⓐ Ⓑ Ⓒ Ⓓ   11. Ⓐ Ⓑ Ⓒ Ⓓ   14. Ⓐ Ⓑ Ⓒ Ⓓ
3. Ⓐ Ⓑ Ⓒ Ⓓ   6. Ⓐ Ⓑ Ⓒ Ⓓ   9. Ⓐ Ⓑ Ⓒ Ⓓ   12. Ⓐ Ⓑ Ⓒ Ⓓ   15. Ⓐ Ⓑ Ⓒ Ⓓ

16.   17.   18.   19.   20.

Section 4

1. Ⓐ Ⓑ Ⓒ Ⓓ   7. Ⓐ Ⓑ Ⓒ Ⓓ   13. Ⓐ Ⓑ Ⓒ Ⓓ   19. Ⓐ Ⓑ Ⓒ Ⓓ   25. Ⓐ Ⓑ Ⓒ Ⓓ
2. Ⓐ Ⓑ Ⓒ Ⓓ   8. Ⓐ Ⓑ Ⓒ Ⓓ   14. Ⓐ Ⓑ Ⓒ Ⓓ   20. Ⓐ Ⓑ Ⓒ Ⓓ   26. Ⓐ Ⓑ Ⓒ Ⓓ
3. Ⓐ Ⓑ Ⓒ Ⓓ   9. Ⓐ Ⓑ Ⓒ Ⓓ   15. Ⓐ Ⓑ Ⓒ Ⓓ   21. Ⓐ Ⓑ Ⓒ Ⓓ   27. Ⓐ Ⓑ Ⓒ Ⓓ
4. Ⓐ Ⓑ Ⓒ Ⓓ   10. Ⓐ Ⓑ Ⓒ Ⓓ   16. Ⓐ Ⓑ Ⓒ Ⓓ   22. Ⓐ Ⓑ Ⓒ Ⓓ   28. Ⓐ Ⓑ Ⓒ Ⓓ
5. Ⓐ Ⓑ Ⓒ Ⓓ   11. Ⓐ Ⓑ Ⓒ Ⓓ   17. Ⓐ Ⓑ Ⓒ Ⓓ   23. Ⓐ Ⓑ Ⓒ Ⓓ   29. Ⓐ Ⓑ Ⓒ Ⓓ
6. Ⓐ Ⓑ Ⓒ Ⓓ   12. Ⓐ Ⓑ Ⓒ Ⓓ   18. Ⓐ Ⓑ Ⓒ Ⓓ   24. Ⓐ Ⓑ Ⓒ Ⓓ   30. Ⓐ Ⓑ Ⓒ Ⓓ

31.   32.   33.   34.   35.

36.   37.   38.

Section 3 — No Calculator

TIME: 25 minutes, 20 questions

Directions

For questions 1-15, solve each problem, choose the best answer from the choices provided, and fill in the corresponding bubble on your answer sheet. Refer to the directions before question 16 on how to enter your answers in the grid. You may use any available space in your test booklet for scratch work.

Notes

- The use of a calculator **is not permitted.**

- All variables and expressions used represent real numbers unless otherwise indicated.

- Figures provided in this test are drawn to scale unless otherwise indicated.

- All figures lie in a plane unless otherwise indicated.

- Unless otherwise indicated, the domain of a given function f is the set of all real numbers x for which $f(x)$ is a real number.

Reference

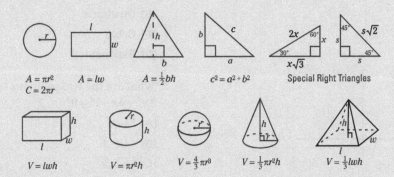

$A = \pi r^2$
$C = 2\pi r$

$A = lw$

$A = \frac{1}{2}bh$

$c^2 = a^2 + b^2$

Special Right Triangles

$V = lwh$

$V = \pi r^2 h$

$V = \frac{4}{3}\pi r^3$

$V = \frac{1}{3}\pi r^2 h$

$V = \frac{1}{3}lwh$

The number of degrees of arc in a circle is 360.

The number of radians of arc in a circle is 2π.

The sum of measures in degrees of the angles of a triangle is 180.

---

1. If $3x + 4 = x$, then $2x + 7 =$

(A) −11

(B) −3

(C) 3

(D) 11

2. If $i = \sqrt{-1}$, what is the sum of $5 + 2i$ and $3 - 5i$?

(A) 8

(B) −15i

(C) $8 - 3i$

(D) $15 - 10i$

3. A balloon resting at a height of 5 meters above the ground is released from its mooring and begins to rise at a rate of 3 meters per second. Which of the following functions gives the height of the balloon above the ground t seconds after being released?

(A) $f(t) = 3t + 5$

(B) $f(t) = 3t - 5$

(C) $f(t) = 5t + 3$

(D) $f(t) = 5t - 3$

GO ON TO NEXT PAGE

4. A bag contains five blue tiles, three red tiles, and one white tile. What is the probability that a tile drawn at random from the bag will be red?

(A) $\frac{1}{3}$

(B) $\frac{2}{3}$

(C) $\frac{4}{9}$

(D) $\frac{5}{9}$

5. Which of the following graphs shows a polynomial function that is both is positive and has a degree that's an odd number?

(A)

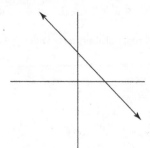

(B)

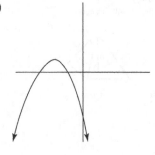

(C)

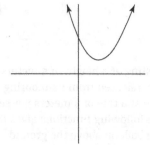

(D)

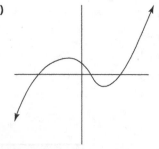

6. Which of the following equations when graphed in the xy-plane has a negative slope and a positive y-intercept?

(A) $3x + 4y = 7$

(B) $6x + y = -2$

(C) $-8x + y = 9$

(D) $-5x - 2y = 10$

x	f(x)
0	a
2	2a
4	4a
6	8a

7. If $a > 1$, then the function $f(x)$, as shown in the table, can best be described as

(A) Linear

(B) Quadratic

(C) Cubic

(D) Exponential

8. What are the two solutions to the equation $3x^2 + 6x - 15 = 0$?

(A) $1 \pm \sqrt{6}$

(B) $-1 \pm \sqrt{6}$

(C) $1 \pm 2\sqrt{2}$

(D) $-1 \pm 2\sqrt{2}$

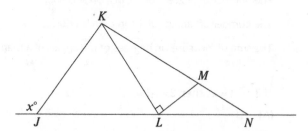

9. In the geometric figure shown here, $JK = KL$, $LM = MN$, and triangle KLM is a right triangle. If $x° = 115°$ then angle LMN measures which of the following?

(A) 130°

(B) 135°

(C) 140°

(D) 145°

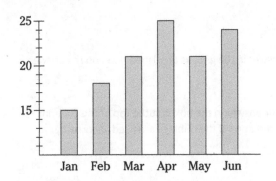

10. This graph shows the number of houses a real estate agency sold in the first six months of last year. In which pair of months are these numbers in a 5:7 ratio?

(A) January and March

(B) February and June

(C) March and April

(D) May and June

11. Which of the following is the equation for a circle that, when graphed on the xy-plane, lies completely within Quadrant 1?

(A) $(x+3)^2 + (y+3)^2 = 4$

(B) $(x-3)^2 + (y-3)^2 = 4$

(C) $(x+3)^2 + (y+3)^2 = 16$

(D) $(x-3)^2 + (y-3)^2 = 16$

12. Raymond owns shares in a stock that are currently worth $10,800. If this amount includes a 35% profit over what he paid for the shares last year, what was the purchase price of the shares?

(A) $6,500

(B) $7,000

(C) $7,500

(D) $8,000

13. Janey wants to cover two rooms in her house with wood flooring. The hickory flooring costs $10.85 per square foot, while the Brazilian walnut flooring costs $16.70 per square foot. Janey ordered a total of 394 square feet of both types of flooring, paying a total of $4,965.20 for it. If H represents the number of square feet of hickory flooring that she purchased, and B represents the number of square feet of Brazilian walnut flooring that she purchased, which of the following systems of equations could be used to find out how much Janey spent on each type of flooring?

(A) $10.85H + 16.70B = 4,965.20$
$H + B = 394$

(B) $10.85H + 16.70B = 394$
$H + B = 4,965.20$

(C) $16.70H + 10.85B = 4,965.20$
$H + B = 394$

(D) $16.70H + 10.85B = 394$
$H + B = 4,965.20$

14. Which of the following expressions is equivalent to $\dfrac{x^2 + 2x - 7}{x - 5}$?

(A) $x + 7 + \dfrac{28}{x-5}$

(B) $x + 7 - \dfrac{28}{x-5}$

(C) $x + 7 + \dfrac{42}{x-5}$

(D) $x + 7 - \dfrac{42}{x-5}$

15. If $a = x^2 - \dfrac{y}{2}$ and $b = x^2 + \dfrac{y}{4}$, what is the value of $4ab$?

(A) $4x^4 + x^2 y - y^2$

(B) $4x^4 - x^2 y - y^2$

(C) $4x^4 + x^2 y - \dfrac{y^2}{2}$

(D) $4x^4 - x^2 y - \dfrac{y^2}{2}$

GO ON TO NEXT PAGE

For questions 16-20, solve the problem and enter your answer in the grid, as described here, on the answer sheet.

1. Although not required, it is suggested that you write your answer in the boxes at the top of the columns to help you fill in the bubbles accurately. You will receive credit only if the bubbles are filled in correctly.

2. Mark no more than one bubble in any column.

3. No question has a negative answer.

4. Some problems may have more than one correct answer. In such cases, grid only one answer.

5. **Mixed numbers** such as $3\frac{1}{2}$ must be gridded as 3.5 or $\frac{7}{2}$. (If 3 1/2 is entered into the grid, it will be interpreted as $\frac{31}{2}$, not $3\frac{1}{2}$.)

6. **Decimal answers:** If you obtain a decimal answer with more digits than the grid can accommodate, it may be either rounded or truncated, but it must fill the entire grid.

---

16. If a line intersects the graph of $y = x^2 - x + 4$ at the point $(0, p)$, then what does p equal?

17. Dianna scored 85, 89, 93, 96, and 100 points on her first 5 chemistry quizzes of the semester. What is the minimum score she must receive on her 6th quiz in order to maintain an average of at least 90 points?

18. If $\sin x° = \cos 2x°$, and $0 < x < 90$, then what does x equal?

19. Lines j and k intersect perpendicularly on the xy-plane when $x = -5$. If the equation of line j is $y = -2x + 3$, then at what y-value does line k intersect the y-axis?

$$3x + 2y = 4$$
$$ax + by = 20$$

20. If the system of linear equations shown here has infinitely many solutions, then what is the value of $a + b$?

DO NOT TURN THE PAGE UNTIL TOLD TO DO SO **STOP** DO NOT RETURN TO A PREVIOUS TEST

Section 4 — Calculator

TIME: 55 minutes, 38 questions

Directions

For questions 1-30, solve each problem, choose the best answer from the choices provided, and fill in the corresponding bubble on your answer sheet. Refer to the directions before question 31 on how to enter your answers in the grid. You may use any available space in your test booklet for scratch work.

Notes

- The use of a calculator **is permitted.**

- All variables and expressions used represent real numbers unless otherwise indicated.

- Figures provided in this test are drawn to scale unless otherwise indicated.

- All figures lie in a plane unless otherwise indicated.

- Unless otherwise indicated, the domain of a given function f is the set of all real numbers x for which $f(x)$ is a real number.

Reference

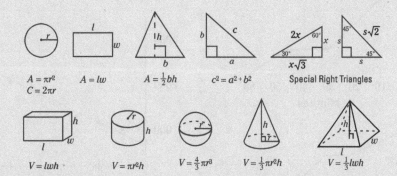

$A = \pi r^2$
$C = 2\pi r$

$A = lw$

$A = \frac{1}{2}bh$

$c^2 = a^2 + b^2$

Special Right Triangles

$V = lwh$

$V = \pi r^2 h$

$V = \frac{4}{3}\pi r^3$

$V = \frac{1}{3}\pi r^2 h$

$V = \frac{1}{3}lwh$

The number of degrees of arc in a circle is 360.

The number of radians of arc in a circle is 2π.

The sum of measures in degrees of the angles of a triangle is 180.

---

1. Isabella opened a bank account with an initial deposit of $600. Every week, she deposits $30 more to the account. If w represents the number of weeks that Isabella has been depositing money into her bank account, and $D(w)$ represents the dollar amount that she has saved, which of the following linear functions models the amount that Isabella's bank account contains?

   (A) $D(w) = 30w + 600$

   (B) $D(w) = 30w - 600$

   (C) $D(w) = 600w + 30$

   (D) $D(w) = 600w - 30$

2. If $2x - 7 = 3(4x + 1)$, then $3x + 4 =$

   (A) -2

   (B) -1

   (C) 1

   (D) 2

GO ON TO NEXT PAGE

3. Leonard took 60 minutes to travel on foot home from the library, a distance of 3 miles. He ran at a constant rate for 10 minutes, then walked more slowly at a constant rate for another 20 minutes. He stopped for 15 minutes to get a slice of pizza, and then walked the rest of the way home. Which of the graphs models Leonard's distance from home during the 60 minutes of his journey?

(A)

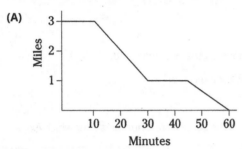

(B)

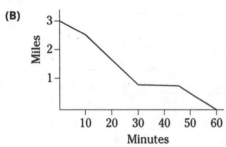

(C)

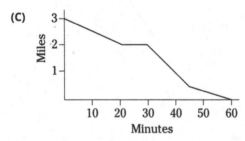

(D)

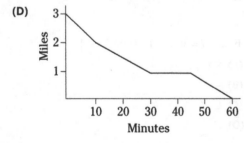

4. At 2,722 feet, the tallest building in the world is the Burj Khalifa in Dubai. In comparison, the Shanghai Tower in China stands at 2,073 feet. Which of the following is the best estimate of the ratio of the Shanghai Tower's height to that of the Burj Khalifa?

(A) 2:3

(B) 3:4

(C) 7:10

(D) 8:10

5. $\dfrac{x^2 y^3}{x^6 y^{-4}} =$

(A) $x^4 y^7$

(B) $\dfrac{x^4}{y}$

(C) $\dfrac{y^7}{x^4}$

(D) $\dfrac{1}{x^4 y}$

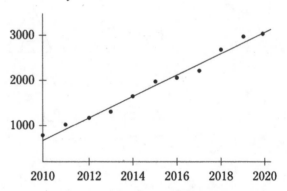

6. This graph provides data on the number of employees at a company from 2010 to 2020, along with the line of best fit for the data shown. Using x as an input variable signifying the number of years after 2010, which of the following is the best approximation for the equation of the line of best fit, as shown in the graph?

(A) $y = 200x + 400$

(B) $y = 200x + 600$

(C) $y = 600x + 200$

(D) $y = 600x + 400$

7. If data set *A* includes 25 values and has a median of 50, what is the maximum number of values in that set that could be greater than 80?

(A) 0

(B) 12

(C) 13

(D) 24

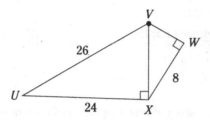

8. This figure includes two right triangles *VXU* and *VWX*. If *UV* = 26, *WX* = 8, and *UX* = 24, then the length of *VW* is

(A) 5

(B) 5.5

(C) 6

(D) 6.5

$$x = \frac{5}{x+4}$$

9. What are the two values of *x* that satisfy this equation?

(A) 1 and 5

(B) 1 and −5

(C) −1 and 5

(D) −1 and −5

10. If $3(4x - 3y) = 6 - y$, what is the value of $3x - 2y$?

(A) $\frac{2}{3}$

(B) $\frac{3}{2}$

(C) $-\frac{2}{3}$

(D) $-\frac{3}{2}$

Questions 11 and 12 refer to the following information:

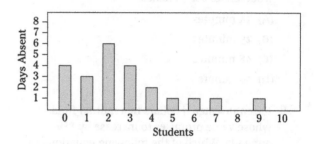

This histogram shows absentee data for the 23 students in Ms. Ciampa's homeroom class for the current school year.

11. If Ms. Ciampa were to pick one student at random from her homeroom class, what is the probability that this student will have been absent <u>at least</u> 4 days this year?

(A) $\frac{2}{23}$

(B) $\frac{4}{23}$

(C) $\frac{6}{23}$

(D) $\frac{8}{23}$

12. What is the median number of absentee days for the entire class?

(A) 2 days

(B) 3 days

(C) 4 days

(D) 5 days

GO ON TO NEXT PAGE

$$-2y \geq -2x + 6$$
$$x + 1 \leq 5$$

13. In the system of linear inequalities shown here, which of the following is true for all values in the solution set?

(A) $y \leq 1$

(B) $y \geq 1$

(C) $y \leq -1$

(D) $y \geq -1$

14. A team of 3 servers can set 10 tables in 8 minutes. Assuming that all of the servers work at the same rate, how long will it take 2 servers to set 20 tables?

(A) 24 minutes

(B) 25 minutes

(C) 48 minutes

(D) 50 minutes

15. Karen invested d dollars in a mutual fund whose value promises to increase by 8% annually. Which of the following equations can be used to calculate $V(x)$ as the value of the mutual fund in x years?

(A) $V(x) = 0.08(d)^x$

(B) $V(x) = 1.08(d)^x$

(C) $V(x) = d(0.08)^x$

(D) $V(x) = d(1.08)^x$

16. The Kingston City Board of Supervisors is trying to determine the popularity of a new proposal to improve the basketball court in the city's public park. They designed a survey about the proposal and asked 100 patrons of a local sports bar to complete it. They found that 86 were in favor of the proposal and 14 were opposed. Which of the following conclusions can be drawn from these survey results?

(A) Exactly 86% of Kingston City residents are currently in favor of the proposal.

(B) Approximately 86% of Kingston City residents are currently in favor of the proposal.

(C) No more than 14% of Kingston City residents are currently opposed to the proposal.

(D) No conclusion can be drawn because the survey's sampling method was flawed.

$$\sqrt{2x + 15} = x$$

17. What are all values of x that satisfy this equation?

I. -3

II. 5

(A) I only

(B) II only

(C) I and II

(D) Neither I nor II

$$5x - 3y = 10$$
$$-4x + 4y = 5$$

18. In the system of linear equations shown here, $3x + 3y =$

(A) 45

(B) 55

(C) 65

(D) 75

19. If $f(x) = 2x + 3$ and $g(x) = 5x - 1$, then $g(f(4)) =$

(A) 41

(B) 46

(C) 50

(D) 54

20. If P percent of 70 equals 8.75, then what percent of P equals 70?

(A) 0.56%

(B) 5.6%

(C) 56%

(D) 560%

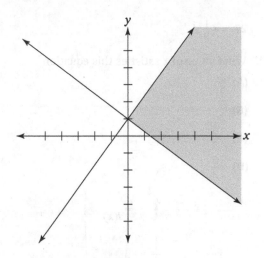

21. Which of the following systems of linear inequalities has a solution set that corresponds to the shaded region in this graph?

(A)
$$y \le -\frac{3}{4}x + 1$$
$$3x - 2y \le -2$$

(B)
$$y \le -\frac{3}{4}x + 1$$
$$3x - 2y \ge -2$$

(C)
$$y \ge -\frac{3}{4}x + 1$$
$$3x - 2y \le -2$$

(D)
$$y \ge -\frac{3}{4}x + 1$$
$$3x - 2y \ge -2$$

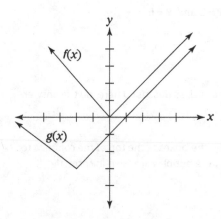

22. This figure shows the graph of $f(x) = |x|$. The graph of $g(x)$ is a transformation that displaces $f(x)$ 2 units to the left and 3 units down. Which of the following is $g(x)$?

(A) $g(x) = |x + 2| + 3$

(B) $g(x) = |x - 2| + 3$

(C) $g(x) = |x + 2| - 3$

(D) $g(x) = |x - 2| - 3$

23. Line m and n have parallel graphs when drawn on the xy-plane. If the equation of line m is $3x - 4y = 12$, and line n includes the point $(1, -2)$, what is the equation of line n?

(A) $y = \frac{3}{4}x + \frac{11}{4}$

(B) $y = \frac{3}{4}x - \frac{11}{4}$

(C) $y = -\frac{3}{4}x + \frac{5}{4}$

(D) $y = -\frac{3}{4}x - \frac{5}{4}$

24. If $i^2 = -1$, what is the result when you multiply $2 + 3i$ by $5 - i$?

(A) $13 + 7i$

(B) $13 + 13i$

(C) $17 + 7i$

(D) $17 + 13i$

	Monthly	Yearly	Total
Gold	146	78	224
Silver	281	93	374
Total	427	171	598

25. Gary's Gym features two membership levels (gold and silver) and two payment options (monthly and yearly). This table shows the breakdown of membership levels and payment options for all 598 current members of the gym. If one gold-level member is picked at random to receive a free personal training course, to the nearest whole percent, what is the probability that this member has chosen the yearly payment option?

(A) 13%

(B) 35%

(C) 46%

(D) 78%

$$(2x - 5)(x + n) = 2x^2 - kx + 30$$

26. If this equation is true for all values of x, then $k + n =$

(A) 9

(B) 10

(C) 11

(D) 12

GO ON TO NEXT PAGE

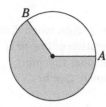

27. In this circle, the shaded area of 40π represents 62.5% of the total area of the circle. Which of the following is the arc length from A to B on the circle?

(A) 4π

(B) 5π

(C) 6π

(D) 8π

$$\frac{x^3 - 3x^2 - 4x + 12}{x^3 - x^2 - 6x}$$

28. Which of the following is equivalent to this rational expression?

(A) $\frac{x+2}{x}$

(B) $\frac{x-2}{x}$

(C) $\frac{x+3}{x}$

(D) $\frac{x-3}{x}$

$$27^x = \left(\frac{1}{9}\right)^{x-1}$$

29. What value of x satisfies this equation?

(A) $\frac{2}{5}$

(B) $\frac{5}{2}$

(C) $-\frac{2}{5}$

(D) $-\frac{5}{2}$

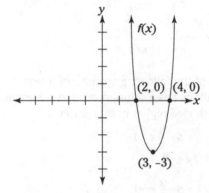

30. In this figure, the function $f(x) = u(x-u)^2 - u$ has roots at $x = 2$ and $x = 4$. If $v = u + 1$ and $g(x) = v(x-v)^2 - v$, then the roots of $g(x)$ are at

(A) $x = 1$ and $x = 3$

(B) $x = 3$ and $x = 5$

(C) $x = 1$ and $x = 5$

(D) $x = 2$ and $x = 6$

Directions

For questions 31-38, solve the problem and enter your answer in the grid, as described here, on the answer sheet.

- Although not required, it is suggested that you write your answer in the boxes at the top of the columns to help you fill in the bubbles accurately. You will receive credit only if the bubbles are filled in correctly.

- Mark no more than one bubble in any column.

- No question has a negative answer.

- Some problems may have more than one correct answer. In such cases, grid only one answer.

- **Mixed numbers** such as $3\frac{1}{2}$ must be gridded as 3.5 or $\frac{7}{2}$. (If 3 1/2 is entered into the grid, it will be interpreted as $\frac{31}{2}$, not $3\frac{1}{2}$.)

- **Decimal answers:** If you obtain a decimal answer with more digits than the grid can accommodate, it may be either rounded or truncated, but it must fill the entire grid.

31. A hectare of land is approximately equal to 2.47 acres. If a farmer owns 15 hectares of land, how many acres do they own, rounded to the nearest whole acre?

32. If $x \geq 0$, for what value of x is $|-7x + 3| + 4$ equal to 15?

$V = 34.375\pi$

$r = 2.5$

33. The cylinder shown here has a volume of 34.375π cubic centimeters. The radius of the can's base is 2.5 centimeters. What is the height of the can?

x	y
0	3
2	9
3	12
4	15

34. This table shows four values for (x,y) for a linear function $y = px + q$. What is the value of $p - q$?

35. An elevator has a maximum capacity of 1,500 pounds. Maxwell wants to use it to help transport a shipment of exercise equipment as quickly as possible. If he needs to transport 200 dumbbells, each of which weighs 40 pounds, what is the minimum number of trips he will require on the elevator to accomplish this task without exceeding its maximum capacity? (Note: Each dumbbell is a single unit that cannot be dismantled into smaller pieces.)

36. In this triangle, $\cos\theta = \dfrac{5\sqrt{146}}{146}$. What is the value of $\tan\theta$?

Questions 37 and 38 refer to the following information.

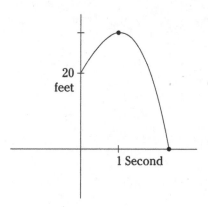

20 feet

1 Second

$h(t) = -16t^2 + vt + 20$

Marci launched a baseball into the air from a height of 20 feet and observed the results. She sketched the graph shown here, noting that the ball reached its highest point exactly 1 second after release. She used the equation shown here to model the height of the ball in feet (h) over time in seconds (t), using v as an unknown quantity equaling the velocity of the ball upon release in feet per seconds.

37. What was the maximum height above the ground, in feet, that the baseball reached?

38. How many seconds was the baseball in the air before it hit the ground?

Chapter 17

Answers and Explanations for Practice SAT Math Test 1

H ere are the answers to the 58 questions from Practice SAT Math Test 1 in Chapter 16. For each question, I provide a detailed explanation showing a possible way to find the correct answer. I recommend reading these explanations to further hone your SAT math skills.

Section 3 — No Calculator

1. **C. 3.** Begin by solving the equation for x:

   $$3x + 4 = x$$
   $$4 = -2x$$
   $$-2 = x$$

   Now, plug -2 for x into $2x + 7$:

   $$2x + 7 = 2(-2) + 7 = -4 + 7 = 3$$

   Therefore, Answer C is correct.

2. **C. 8 − 3i.** To add complex numbers, add the real parts $5 + 3 = 8$ and the imaginary parts $2i + (-5i) = -3i$, so:

   $$(5 + 2i) + (3 - 5i) = 8 - 3i$$

   Therefore, Answer C is correct.

3. **A. $f(t) = 3t + 5$.** The height of the balloon is modeled by the linear function $y = mx + b$. The starting height of the balloon is 5 meters, so this is the y-intercept b. The rate of change is an increase of 3 meters per second, so this is the slope m. Thus, $y = 3x + 5$. In this case, you substitute t for x and $f(t)$ for y, so $f(t) = 3t + 5$. Therefore, Answer A is correct.

4. **A.** $\frac{1}{3}$. Use the formula for probability:

$$\text{Probability} = \frac{\text{Target Outcomes}}{\text{Total Outcomes}}$$

Here, the total number of outcomes equals the number of tiles in the bag, which is 9. And the target outcomes equal the number of red tiles, which is 3:

$$= \frac{3}{9} = \frac{1}{3}$$

Therefore, Answer A is correct.

5. **D.** A positive polynomial function has positive end behavior in the positive direction — that is, as x increases, y increases — which rules out Answers A and B. An odd polynomial function has opposite end behaviors in the positive and negative directions — that is, it explodes to positive infinity in one direction and negative infinity in the other — which rules out Answer C. Therefore, Answer D is correct.

6. **A.** $3x + 4y = 7$. Isolate y to change each equation from standard form to slope-intercept form:

$$3x + 4y = 7$$
$$4y = -3x + 7$$
$$y = -\frac{3}{4}x + \frac{7}{4}$$

This equation has a slope of $-\frac{3}{4}$, which is negative, and a y-intercept of $\frac{7}{4}$, which is positive. Each of the other three answers has a different combination of slope and y-intercept, so they are all incorrect. Therefore, Answer A is correct.

7. **D. Exponential.** As the x-values increase in a linear fashion, the y-values increase by a factor of 2. Thus, the function is exponential, so Answer D is correct.

8. **B.** $-1 \pm \sqrt{6}$. Begin by dividing the equation by 3:

$$3x^2 + 6x - 15 = 0$$
$$x^2 + 2x - 5 = 0$$

Next, plug in 1 for a, 2 for b, and -5 for c into the quadratic equation:

$$x = \frac{-b \pm \sqrt{b^2 - 4ac}}{2a} = \frac{-2 \pm \sqrt{2^2 - 4(1)(-5)}}{2(1)}$$

Simplify:

$$= \frac{-2 \pm \sqrt{4 + 20}}{2}$$
$$= \frac{-2 \pm \sqrt{24}}{2}$$

This result still doesn't look like any of the four answers, but it can be simplified by splitting the radical into two factors:

$$= \frac{-2 \pm \sqrt{4}\sqrt{6}}{2}$$
$$= \frac{-2 \pm 2\sqrt{6}}{2}$$

Now, you can reduce the numerator and denominator by a factor of 2 as follows:

$$= \frac{2(-1 \pm \sqrt{6})}{2}$$
$$= -1 \pm \sqrt{6}$$

Therefore, Answer B is correct.

9. **A. 130°.** The problem states that $x° = 115°$, so angle *KJL* measures $180° - 115° = 65°$. $JK = KL$, so triangle *JKL* is isosceles; thus, angle *KLJ* also measures 65°. Angle *MLK* measures 90°, so angle *NLM* measures $180° - 65° - 90° = 25°$. $LM = MN$, so triangle *LMN* is isosceles; thus, angle *MNL* also measures 25°. Therefore, angle *LMN* measures $180° - 25° - 25° = 130°$, so Answer A is correct.

10. **A. January and March.** The graph shows January sales at 15 houses, February at 18, March at 21, April at 24, May at 21, and June at 24. Thus, sales for January and March are in a 15:21 ratio, which simplifies to 5:7, so Answer A is correct.

11. **B. $(x-3)^2 + (y-3)^2 = 4$.** Answers A and C are both equations for circles centered at the point $(-3,-3)$, which rules out both of these answers. Answers B and D are both equations for circles centered at the point $(3,3)$. However, Answer D is an equation for a circle with a radius of 4, so this circle doesn't stay entirely in Quadrant 1. Answer B is an equation for a circle with radius 2, so this circle fits entirely in Quadrant 1. Thus, Answer B is correct.

12. **D. $8,000.** Let x be the purchase price of the shares in dollars. Then 135% of x equals 10,800, so:

$$1.35x = 10,800$$

Solve by dividing each side by 1.35:

$$\frac{1.35x}{1.35} = \frac{10,800}{1.35}$$
$$x = 8,000$$

Thus, Raymond paid $8,000, so Answer D is correct.

13. **A. 10.85H + 16.70B = 4,965.20, H + B = 394.** *H* and *B* stand, respectively, for the number of square feet of hickory and Brazilian walnut flooring that Janey purchased. This total number of square feet $(H + B)$ adds up to 394, so you can rule out Answers B and D. Next, note that *H* (hickory) costs $10.85 and that *B* (Brazilian walnut) costs $16.70, so these values in the first equation need to be paired correctly. This rules out Answer C, so Answer A is correct.

14. **A. $x + 7 + \dfrac{28}{x-5}$.** To solve this problem, divide $x^2 + 2x - 7$ by $x - 5$ using synthetic division:

$$
\begin{array}{r|rrr}
 & 1 & 2 & -7 \\
5 \downarrow & & 5 & 35 \\
\hline
 & 1 & 7 & 28 \\
\end{array}
$$

Thus, the original expression equals $x + 7 + \dfrac{28}{x-5}$, so Answer A is correct.

15. **D.** $4x^4 - x^2y - \dfrac{y^2}{2}$. Begin by substituting $x^2 - \dfrac{y}{2}$ for a and $x^2 + \dfrac{y}{4}$ for b into the expression $4ab$:

$$4ab = 4(x^2 - \frac{y}{2})(x^2 + \frac{y}{4})$$

Now distribute 4 into the first set of parentheses:

$$= (4x^2 - 2y)(x^2 + \frac{y}{4})$$

Next, FOIL the contents of these two sets of parentheses:

$$= 4x^4 + x^2y - 2x^2y - \frac{y^2}{2}$$

To finish, combine like terms:

$$= 4x^4 - x^2y - \frac{y^2}{2}$$

Therefore, Answer D is correct.

16. **4.** To find p, plug 0 for x into the equation and solve for y:

$$y = x^2 - x + 4$$
$$y = 0^2 - 0 + 4$$
$$y = 4$$

17. **77.** To score an average of 90 points on 6 quizzes, she needs to score a total of $90 \times 6 = 540$ points. Her current number of points is $85 + 89 + 93 + 96 + 100 = 463$. Thus, to score a total of 540 points, the number of points she needs on her 6th quiz is $540 - 463 = 77$.

18. **30.** The trig identity $\sin x° = \cos(90 - x)°$ holds for all values of x. Thus, because $\sin x° = \cos 2x°$, you can create and solve the following equation:

$$2x = 90 - x$$
$$3x = 90$$
$$x = 30$$

19. **31/2 or 15.5.** Lines j and k intersect when $x = -5$, so to find the y-value of this point, plug -5 for x into the equation for line j:

$$y = -2x + 3 = -2(-5) + 3 = 10 + 3 = 13$$

Thus, lines j and k intersect at $(-5, 13)$. The slope of line j is -2. Lines j and k intersect perpendicularly, so the slope of line k is $\dfrac{1}{2}$. Thus, the equation for line k is:

$$y = \frac{1}{2}x + b$$

To find b, plug in -5 for x and 13 for y and solve:

$$13 = \frac{1}{2}(-5) + b$$
$$13 = -2.5 + b$$
$$15.5 = b$$

This value is the y-intercept of line k, so line k crosses the y-axis at 15.5, which also equals $\frac{31}{2}$.

20. **25.** A system of linear equations has infinitely many solutions only when the x-term, y-term, and constant are all in proportion to each other. In this system, the constant in the second equation is 5 times that in the first. Thus, a must be 5 times 3, so $a = 15$. Similarly, b must be 5 times 2, so $b = 10$. Therefore, $a + b = 25$.

Section 4 — Calculator

1. **A. $D(w) = 30w + 600$.** The linear function $y = mx + b$ models the amount of money in the bank account. The initial value of \$600 is the y-intercept b, and the \$30 incremental rate of increase is the slope m. Thus, the linear function $y = 30x + 600$ models this amount. Substituting w as the input value x and $D(w)$ as the output value y results in the function $D(w) = 30w + 600$, so Answer A is correct.

2. **C. 1.** Begin by solving the equation for x:

$$2x - 7 = 3(4x + 1)$$
$$2x - 7 = 12x + 3$$
$$-7 = 10x + 3$$
$$-10 = 10x$$
$$-1 = x$$

Now, plug in -1 for x into the expression $3x + 4$:

$$3x + 4 = 3(-1) + 4 = -3 + 4 = 1$$

Therefore, Answer C is correct.

3. **D.**

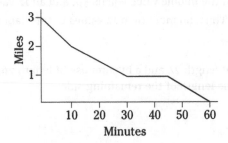

Leonard ran for only the first 10 minutes of his journey, so this should be represented by a steep downward slope. He walked for the next 20 minutes, so this should be represented by a more shallow downward slope. He stopped for the next 15 minutes, so this should be represented by a horizontal line. Finally, he walked for the last 15 minutes, so this should be represented by a shallow downward slope. All of these features are shown in Answer D, so this is the correct answer.

4. **B. 3:4.** To find the ratio of the Shanghai Tower's height to that of the Burj Khalifa, divide 2,073 by 2,722:

$$\frac{2,073}{2,722} \approx 0.76$$

This decimal value is closer to the fraction $\frac{3}{4} = 0.75$ than to any of the other answers, so Answer B is correct.

5. **C. $\frac{y^7}{x^4}$.** To begin, divide the x and y terms by subtracting the exponent of the numerator from the exponent of the denominator:

$$\frac{x^2 y^3}{x^6 y^{-4}} = x^{2-6} y^{3-(-4)} = x^{-4} y^7$$

To complete the problem, move the x term to the denominator, changing its exponent from negative to positive:

$$= \frac{y^7}{x^4}$$

Therefore, Answer C is correct.

6. **B. $y = 200x + 600$.** The line of best fit crosses the y-intercept at approximately 600, so this is the y-intercept b for the function. It also rises by about 200 every year, so this is the slope m. Thus, the function that approximates the line of best fit is:

$$y = mx + b = 200x + 600$$

Therefore, Answer B is correct.

7. **B. 12.** To find the median of a data set with 25 values, arrange these values left to right from least to greatest:

X X X X X X X X X X X X 50 X X X X X X X X X X X X

If the median of this data set is 50, then the middle value equals 50, and all 12 values to the left of this are less than or equal to 50. Thus, no more than 12 values in this data set could be greater than 80, so Answer B is correct.

8. **C. 6.** The right triangle VXU has a side of length 24 and a hypotenuse of length 26, so you can use the Pythagorean Theorem to find the length of the remaining side:

$$a^2 + b^2 = c^2$$
$$a^2 + 24^2 = 26^2$$
$$a^2 + 576 = 676$$
$$a^2 = 100$$
$$a = 10$$

So VX has a length of 10. Thus, VWX is a right triangle with a side of length 8 and a hypotenuse of 10, making this a 3-4-5 right triangle with lengths doubled, so VW has a length of 6. (You can also use the Pythagorean Theorem to find this length.) Therefore, Answer C is correct.

9. **B. 1 and −5.** Start by solving the equation:

$$x = \frac{5}{x+4}$$
$$x(x+4) = 5$$
$$x^2 + 4x = 5$$
$$x^2 + 4x - 5 = 0$$
$$(x+5)(x-1) = 0$$

Now, break the equation into two separate equations and solve separately:

$$x + 5 = 0 \qquad\qquad x - 1 = 0$$
$$x = -5 \qquad\qquad x = 1$$

Thus, Answer B is correct.

10. **B. $\frac{3}{2}$.** Begin by distributing and combining like terms:

$$3(4x - 3y) = 6 - y$$
$$12x - 9y = 6 - y$$
$$12x - 8y = 6$$

Now, notice that the left side of the equation is 4 times the expression $3x - 2y$. This means you can divide both sides of this equation by 4:

$$3x - 2y = \frac{6}{4} = \frac{3}{2}$$

Therefore, Answer B is correct.

11. **C. $\frac{6}{23}$.** Out of a pool of 23 students, 6 students have been absent at least 4 days this year. Place this information into the formula for probability:

$$\text{Probability} = \frac{\text{Target outcomes}}{\text{Total outcomes}} = \frac{6}{23}$$

Therefore, Answer C is correct.

12. **A. 2 days.** To begin, transcribe the data from the histogram into a list of 23 data points, each showing the number of absent days for that student:

0, 0, 0, 0, 1, 1, 1, 2, 2, 2, 2, **2**, 2, 3, 3, 3, 3, 4, 4, 5, 6, 7, 9

This list shows that 4 students were absent 0 days, 3 students were absent 1 day, and so forth. The middle number in this list is 2, so this is the median number of absent days for the class; therefore, Answer A is correct.

13. **A. $y \leq 1$.** To isolate y in the first inequality, divide both sides by −1, flipping the inequality sign from $\geq$ to $\leq$, and then simplify:

$$\frac{-2y}{-2} \leq \frac{-2x+6}{-2}$$
$$y \leq x - 3$$

Next, to make the left side of the second inequality equal to $x - 3$, add 5 to both sides of the second inequality:

$$x + 1 \leq 5$$
$$x - 3 \leq 1$$

Now, the two inequalities both have $\leq$, so you can arrange them as follows:

$$y \leq x - 3 \leq 1$$

This inequality implies $y \leq 1$, so Answer A is correct.

14. **A. 24 minutes.** A team of 3 servers can set 10 tables in 8 minutes, so the same team can set 20 tables in $8 \times 2 = 16$ minutes. Thus, only 1 server would take $16 \times 3 = 48$ minutes to set 20 tables. Thus, 2 servers would take $48 \div 2 = 24$ minutes to set 20 tables, so Answer A is correct.

15. **D. $V(x) = d(1.08)^x$.** To calculate a yearly percent increase of $8\% = 0.08$ as an exponential function, the base of the exponent must equal 1.08. Therefore, Answer D is correct.

16. **D. No conclusion can be drawn because the survey's sampling method was flawed.** The survey was given to patrons of a sports bar whose interest in sports, such as basketball, might incline them to be more favorable to the proposal than other city residents. Therefore, the survey's sampling method was flawed, so Answer D is correct.

17. **B. II only.** Begin by substituting -3 for x into the equation:

$$\sqrt{2(-3) + 15} = -3$$

This equation must be false, because the radical on the left side of the equation must equal a non-negative number. This rules out Answers A and C. Now, substitute 5 for x into the equation:

$$\sqrt{2(5) + 15} = 5$$
$$\sqrt{10 + 15} = 5$$
$$\sqrt{25} = 5$$
$$5 = 5$$

Thus, 5 satisfies the equation, so Answer B is correct.

18. **A. 45.** Begin by adding the two equations together:

$$x + y = 15$$

Now, multiply the resulting equation by 3:

$$3x + 3y = 45$$

Therefore, Answer A is correct.

19. **D. 54.** To begin, evaluate $f(4)$:

$$f(4) = 2(4) + 3 = 11$$

Now, plug in 11 for $f(4)$ into $g(f(4)) =$

$$g(f(4)) = g(11)$$

To complete the problem, evaluate $g(11)$:

$$= 5(11) - 1 = 55 - 1 = 54$$

20. **D. 560%.** To begin, set up the statement "P percent of 70 equals 8.75" as an equation and solve for P:

$$P(0.01)(70) = 8.75$$
$$0.7P = 8.75$$
$$P = 12.5$$

Now, set up the statement "what percent of 12.5 equals 70" using x to represent the percentage:

$$x(0.01)(12.5) = 70$$
$$0.125x = 70$$
$$x = 560$$

Therefore, Answer D is correct.

21. **D.** $\begin{aligned} y \geq -\frac{3}{4}x + 1 \\ 3x - 2y \geq -2 \end{aligned}$. The line representing the equation $y = -\frac{3}{4}x + 1$ has a negative slope on the xy-plane. The shaded region on the graph is above this line, representing the inequality $y \geq -\frac{3}{4}x + 1$, which rules out Answers A and B. The other line represents the equation $3x - 2y = -2$. The shaded region on the graph is below this line, so the corresponding inequality in slope-intercept form must include the less-than-or-equal-to inequality ($\leq$). Here's how you change each of the two inequalities in Answers C and D to slope-intercept form:

$$\begin{aligned} 3x - 2y &\leq -2 \\ -2y &\leq -3x - 2 \\ y &\geq \frac{3}{2}x + 1 \end{aligned} \qquad \begin{aligned} 3x - 2y &\geq -2 \\ -2y &\geq -3x - 2 \\ y &\leq \frac{3}{2}x + 1 \end{aligned}$$

Note that the inequality $3x - 2y \geq -2$ becomes $y \leq \frac{3}{2}x + 1$, so this inequality is correct. Therefore, Answer D is correct.

22. **C. $g(x) = |x + 2| - 3$.** The horizontal transformation $f(x + 2)$ moves $f(x)$ 2 units to the right. The vertical transformation $f(x) - 3$ moves $f(x)$ 3 units down. Combine these two transformations as follows:

$$g(x) = f(x + 2) - 3 = |x + 2| - 3$$

Therefore, Answer C is correct.

23. **B. $y = \frac{3}{4}x - \frac{11}{4}$.** To begin, change the equation of line m to slope-intercept form:

$$\begin{aligned} 3x - 4y &= 12 \\ -4y &= -3x + 12 \\ y &= \frac{3}{4}x + 12 \end{aligned}$$

Line n is parallel to line m, so line n also has a slope of $\frac{3}{4}$. Thus, the equation for line m is:

$$y = \frac{3}{4}x + b$$

This line includes the point $(1, -2)$, so plug these values into the equation and solve for b:

$$-2 = \frac{3}{4}(1) + b$$

$$-2 = \frac{3}{4} + b$$

$$-2 - \frac{3}{4} = b$$

$$-\frac{11}{4} = b$$

Thus, the equation for line n is:

$$y = \frac{3}{4}x - \frac{11}{4}$$

Therefore, Answer B is correct.

24. **B. $13 + 13i$.** To begin, multiply the two values by FOILing:

$$(2 + 3i)(5 - i) = 10 - 2i + 15i - 3i^2 = 10 + 13i - 3i^2$$

Next, substitute -1 for i^2 in the resulting expression:

$$= 10 + 13i - 3(-1)$$

Simplify:

$$= 10 + 13i + 3 = 13 + 13i$$

Therefore, Answer B is correct.

25. **B. 35%.** To find the conditional probability that a person chosen from the pool of 224 gold-level members will be among the 78 who have chosen the yearly payment option, calculate as follows using the formula for probability:

$$\text{Probability} = \frac{\text{Target outcomes}}{\text{Total outcomes}} = \frac{78}{224} \approx 0.348 \approx 35\%$$

Therefore, Answer B is correct.

26. **C. 11.** Begin by simplifying the left side of the equation:

$$(2x - 5)(x + n) = 2x^2 - kx + 30$$

$$(2x - 5)(x + n) = 2x^2 - kx + 30$$

$$2x^2 + 2nx - 5x - 5n = 2x^2 - kx + 30$$

This equation is true for all values of x, so the constant terms $-5n$ and 30 are equivalent. Thus:

$$-5n = 30$$

$$n = -6$$

Plug this value back into the equation and simplify:

$$2x^2 + 2(-6)x - 5x - 5(-6) = 2x^2 - kx + 30$$
$$2x^2 - 12x - 5x + 30 = 2x^2 - kx + 30$$
$$2x^2 - 17x + 30 = 2x^2 - kx + 30$$
$$-17x = -kx$$
$$17 = k$$

Therefore, $k + n = 17 + (-6) = 11$, so Answer C is correct.

27. **C. 6π.** The shaded area of the circle is 62.5% of the circle, which is $\frac{5}{8}$ of the circle. Thus, to find the total area of the circle, multiply as follows:

$$40\pi \times \frac{8}{5} = 64\pi$$

Now, use the area formula for circles to find the radius of the circle:

$$A = \pi r^2$$
$$64\pi = \pi r^2$$
$$64 = r^2$$
$$8 = r$$

Thus, the circle has a radius of 8. The angle between A and B is

$$\frac{3}{8} \times 360° = 135°$$

To change this value into radians, use the conversion factor $\frac{\pi}{180°}$:

$$135° \times \frac{\pi}{180°} = \frac{3\pi}{4}$$

To find the arc length from A to B, use the arc length formula:

$$\text{Arc length} = \text{radius} \times \text{radians} = \frac{3\pi}{4} \times 8 = 6\pi$$

Therefore, Answer C is correct.

28. **B. $\frac{x-2}{x}$.** To begin, factor the cubic expression in the numerator by grouping:

$$\frac{x^3 - 3x^2 - 4x + 12}{x^3 - x^2 - 6x} = \frac{x^2(x-3) - 4(x-3)}{x^3 - x^2 - 6x} = \frac{(x^2 - 4)(x-3)}{x^3 - x^2 - 6x}$$

Next, factor $(x^2 - 4)$ in the numerator as the difference of squares:

$$= \frac{(x+2)(x-2)(x-3)}{x^3 - x^2 - 6x}$$

Now, factor out x from the cubic expression in the denominator and then factor the resulting quadratic expression:

$$= \frac{(x+2)(x-2)(x-3)}{x^3 - x^2 - 6x} = \frac{(x+2)(x-2)(x-3)}{x(x^2 - x - 6)} = \frac{(x+2)(x-2)(x-3)}{x(x+2)(x-3)}$$

To complete the problem, cancel out factors of $x+2$ and $x-3$ in both the numerator and denominator:

$$= \frac{x-2}{x}$$

Therefore, Answer B is correct.

29. **A.** $\frac{2}{5}$. To begin, rewrite the two bases in the equation as powers of 3, using $27 = 3^3$ and $\frac{1}{9} = 3^{-2}$ as follows:

$$27^x = (\tfrac{1}{9})^{x-1}$$
$$(3^3)^x = (3^{-2})^{x-1}$$

Next, remove both sets of parentheses, in each case by multiplying the pair of exponents:

$$3^{3x} = 3^{-2(x-1)}$$

Now, because the bases are equal, you can drop them and set the exponents equal to each other:

$$3x = -2(x-1)$$

Solve for x:

$$3x = -2x + 2$$
$$5x = 2$$
$$x = \frac{2}{5}$$

30. **B.** $x = 3$ **and** $x = 5$. The function $f(x) = u(x-u)^2 - u$ is in vertex form, which is $f(x) = a(x-h)^2 + k$. Its vertex is at $(h,k) = (3,-3)$, with a stretch factor of $a = 3$, so $u = 3$. Thus, $v = 4$, so $g(x) = 4(x-4)^2 - 4$. The vertex of this function is $(h,k) = (4,-4)$, and its vertical stretch factor a is also 4, so if you graph this function, you can find the roots at $x = 3$ and $x = 5$, as shown in this figure:

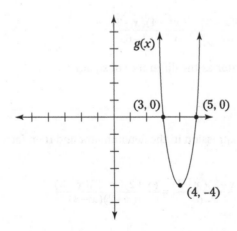

Another way to solve this problem is to change $g(x) = 4(x-4)^2 - 4$ to standard form as follows:

$$g(x) = 4(x^2 - 8x + 16) - 4$$
$$= 4x^2 - 32x + 64 - 4$$
$$= 4x^2 - 32x + 60$$

Next, set this equation to 0 and factor to solve for x:

$$4x^2 - 32x + 60 = 0$$
$$x^2 - 8x + 15 = 0$$
$$(x-3)(x-5) = 0$$

So the roots of this equation are at $x = 3$ and $x = 5$. Therefore, Answer B is correct.

31. **37 acres.** Use the conversion factor $\dfrac{2.47 \text{ acres}}{1 \text{ hectare}}$ to convert 15 acres to hectares:

$$15 \text{ hectares} \times \frac{2.47 \text{ acres}}{1 \text{ hectare}} = 37.05 \text{ acres} \approx 37 \text{ acres}$$

Therefore, the correct answer is 37.

32. **2.** To begin, set up the following equation:

$$|-7x + 3| + 4 = 15$$

Next, subtract 4 from both sides:

$$|-7x + 3| = 11$$

Now, break the equation into two equations and solve separately:

$$-7x + 3 = 11 \qquad\qquad -7x + 3 = -11$$
$$-7x = 8 \qquad\qquad\qquad -7x = -14$$
$$x = -\frac{8}{7} \qquad\qquad\qquad x = 2$$

Because $x \geq 0$, the correct answer is 2.

33. $\dfrac{11}{2}$ **or 5.5.** Plug in 34.375π for V and 2.5 for r into the formula for the volume of a cylinder, and solve for h:

$$V = \pi r^2 h$$
$$34.375\pi = \pi(2.5)^2 h$$
$$34.375 = (2.5)^2 h$$
$$34.375 = 6.25h$$
$$5.5 = h$$

Therefore, this answer is 5.5 or $\dfrac{11}{2}$.

34. **0.** To begin, plug in 0 for x and 3 for y into the function and solve for q:

$$y = px + q$$
$$3 = 0x + q$$
$$3 = q$$

Thus, the function is $y = px + 3$. Now, plug in 2 for x and 9 for y into this function and solve for p:

$$y = px + 3$$
$$9 = 2p + 3$$
$$6 = 2p$$
$$3 = p$$

Therefore, $p - q = 3 - 3 = 0$, so the correct answer is 0.

35. **6.** The elevator has a maximum capacity of 1,500 pounds, so Maxwell can transport at most $1,500 \div 40 = 37.5$ dumbbells. However, because each dumbbell cannot be dismantled, the elevator can hold a maximum of 37 dumbbells. Thus, in 5 trips on the elevator, Maxwell can transport $37 \times 5 = 185$ dumbbells. Therefore, he will require a 6th trip on the elevator to transport the remaining 15 dumbbells, so the answer is 6.

36. $\frac{11}{5}$ **or 2.2.** The value $\frac{5\sqrt{146}}{146}$ is the rationalized form of $\frac{5}{\sqrt{146}}$ because:

$$\frac{5}{\sqrt{146}} = \frac{5}{\sqrt{146}} \cdot \frac{\sqrt{146}}{\sqrt{146}} = \frac{5\sqrt{146}}{146}$$

Thus, $\cos\theta = \frac{5}{\sqrt{146}}$, so the adjacent side and hypotenuse of this triangle are in a ratio of $5 : \sqrt{146}$. Thus, you can use the Pythagorean Theorem to calculate the relative length of the opposite side of the triangle as follows:

$$a^2 + b^2 = c^2$$
$$5^2 + b^2 = \sqrt{146}^2$$
$$25 + b^2 = 146$$
$$b^2 = 121$$
$$b = 11$$

Thus, the opposite and adjacent sides of the triangle are in a ratio of $11 : 5$, so

$$\tan\theta = \frac{11}{5} = 2.2$$

Therefore, the answer is $\frac{11}{5}$ or 2.2.

37. **36.** The baseball reached its maximum height 1 second after it was projected, so the t-coordinate of the vertex is 1. Thus, the axis of symmetry is also 1. Plug in 1 for t, -16 for a, and v for b into the formula for the axis of symmetry as follows:

$$t = -\frac{b}{2a}$$
$$1 = -\frac{v}{2(-16)}$$
$$1 = \frac{v}{32}$$
$$32 = v$$

Thus, you can plug in 32 for v into the equation:

$$h(t) = -16t^2 + 32t + 20$$

To find the height of the baseball 1 second after it was projected, plug in 1 for t:

$$h(1) = -16(1)^2 + 32(1) + 20 = -16 + 32 + 20 = 36$$

Therefore, the correct answer is 36.

38. $\frac{5}{2}$ or 2.5. Begin by setting the function that you created in question 37 equal to 0:

$$0 = -16t^2 + 32t + 20$$

Divide by 4 to simplify:

$$0 = -4t^2 + 8t + 5$$

This quadratic equation can be solved by factoring:

$$0 = -4t^2 + 10t - 2t + 5$$
$$0 = 2t(-2t + 5) + 1(-2t + 5)$$
$$0 = (2t + 1)(-2t + 5)$$

Finish solving by setting both factors equal to 0 and solving separately:

$$0 = 2t + 1 \qquad\qquad 0 = -2t + 5$$
$$-2t = 1 \qquad\qquad 2t = 5$$
$$t = -\frac{1}{2} \qquad\qquad t = \frac{5}{2}$$

The negative solution $-\frac{1}{2}$ is extraneous to this problem, but the positive solution $\frac{5}{2}$ gives the time when the ball hits the ground, so the answer is $\frac{5}{2}$ or 2.5.

Chapter **18**

Practice SAT Math Test 2

Second chances happen so rarely in life. So, here's your second chance to test your SAT math skills. This chapter includes two complete SAT Math Test sections, just like on the actual SAT. Work through each section in the time stated at the top of that section.

Remember that you may **not** use a calculator for the first math section, but you may use one for the second. But to simulate SAT conditions more closely, tear out and use the answer sheet provided on the next page.

When you're done, turn to Chapter 19 for answers and complete explanations.

Good luck!

Answer Sheet for Practice SAT Math Test 2

Use the ovals and grid-ins provided with this practice exam to record your answers.

Section 3

1. Ⓐ Ⓑ Ⓒ Ⓓ   4. Ⓐ Ⓑ Ⓒ Ⓓ   7. Ⓐ Ⓑ Ⓒ Ⓓ   10. Ⓐ Ⓑ Ⓒ Ⓓ   13. Ⓐ Ⓑ Ⓒ Ⓓ
2. Ⓐ Ⓑ Ⓒ Ⓓ   5. Ⓐ Ⓑ Ⓒ Ⓓ   8. Ⓐ Ⓑ Ⓒ Ⓓ   11. Ⓐ Ⓑ Ⓒ Ⓓ   14. Ⓐ Ⓑ Ⓒ Ⓓ
3. Ⓐ Ⓑ Ⓒ Ⓓ   6. Ⓐ Ⓑ Ⓒ Ⓓ   9. Ⓐ Ⓑ Ⓒ Ⓓ   12. Ⓐ Ⓑ Ⓒ Ⓓ   15. Ⓐ Ⓑ Ⓒ Ⓓ

16.   17.   18.   19.   20.

Section 4

1. Ⓐ Ⓑ Ⓒ Ⓓ   7. Ⓐ Ⓑ Ⓒ Ⓓ   13. Ⓐ Ⓑ Ⓒ Ⓓ   19. Ⓐ Ⓑ Ⓒ Ⓓ   25. Ⓐ Ⓑ Ⓒ Ⓓ
2. Ⓐ Ⓑ Ⓒ Ⓓ   8. Ⓐ Ⓑ Ⓒ Ⓓ   14. Ⓐ Ⓑ Ⓒ Ⓓ   20. Ⓐ Ⓑ Ⓒ Ⓓ   26. Ⓐ Ⓑ Ⓒ Ⓓ
3. Ⓐ Ⓑ Ⓒ Ⓓ   9. Ⓐ Ⓑ Ⓒ Ⓓ   15. Ⓐ Ⓑ Ⓒ Ⓓ   21. Ⓐ Ⓑ Ⓒ Ⓓ   27. Ⓐ Ⓑ Ⓒ Ⓓ
4. Ⓐ Ⓑ Ⓒ Ⓓ   10. Ⓐ Ⓑ Ⓒ Ⓓ   16. Ⓐ Ⓑ Ⓒ Ⓓ   22. Ⓐ Ⓑ Ⓒ Ⓓ   28. Ⓐ Ⓑ Ⓒ Ⓓ
5. Ⓐ Ⓑ Ⓒ Ⓓ   11. Ⓐ Ⓑ Ⓒ Ⓓ   17. Ⓐ Ⓑ Ⓒ Ⓓ   23. Ⓐ Ⓑ Ⓒ Ⓓ   29. Ⓐ Ⓑ Ⓒ Ⓓ
6. Ⓐ Ⓑ Ⓒ Ⓓ   12. Ⓐ Ⓑ Ⓒ Ⓓ   18. Ⓐ Ⓑ Ⓒ Ⓓ   24. Ⓐ Ⓑ Ⓒ Ⓓ   30. Ⓐ Ⓑ Ⓒ Ⓓ

31.   32.   33.   34.   35.

36.   37.   38.

Section 3 — No Calculator

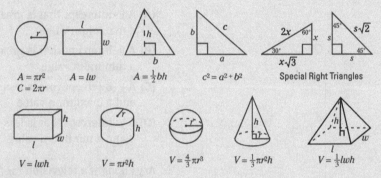
GO ON TO NEXT PAGE

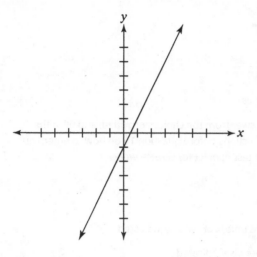

July	August	September	October
$1,700	$2,400	$3,100	$3,800

4. Victoria opened her savings account earlier this year with an initial deposit of s dollars, and every month after that has deposited t dollars to the account. Her bank balance during four months this year is shown in the table. If $s < t$ then in what month did Victoria open her account?

(A) April

(B) May

(C) June

(D) July

$$f(x) = -3x^2 + 14x + 5$$

5. This quadratic function has

(A) A y-intercept that is greater than 0 and a maximum value.

(B) A y-intercept that is greater than 0 and a minimum value.

(C) A y-intercept that is less than 0 and a maximum value.

(D) A y-intercept that is less than 0 and a minimum value.

6. Arya bought a television for $910. If this price reflected a 35% discount from the normal retail price, for what price did the television customarily sell?

(A) $1,228.50

(B) $1,300.00

(C) $1,400.00

(D) $1,428.50

1. Which of the following is the equation of the graph in the xy-plane shown here?

(A) $y = x + 2$

(B) $y = x - 2$

(C) $y = 2x + 1$

(D) $y = 2x - 1$

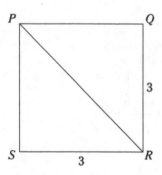

2. In this figure, square $PQRS$ has sides of length 3. What is the length of the diagonal line PR?

(A) $2\sqrt{2}$

(B) $2\sqrt{3}$

(C) $3\sqrt{2}$

(D) $3\sqrt{3}$

$$6xy^5 - 10x^2y^3 + 12x^3y^2$$

3. Which of the following is equivalent to this expression?

(A) $2xy^2(3y^3 - 5xy + 6x^2)$

(B) $2xy^2(3xy^3 - 5xy + 6x^2)$

(C) $2xy^2(3y^3 - 5xy + 6x^2y)$

(D) $2xy^2(3xy^3 - 5xy + 6x^2y)$

Projected 2035 populations of the world's five most populous countries.

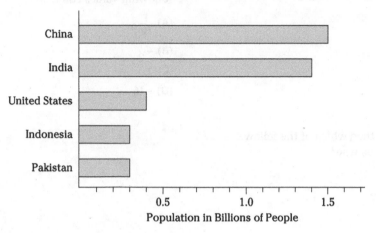

Population in Billions of People

7. According to the graph, the ratio of the projected population of China to that of Indonesia is approximately

(A) 4 to 1

(B) 5 to 1

(C) 5 to 2

(D) 7 to 2

8. According to the graph, what will be the combined population of the four countries other than Indonesia?

(A) 360,000,000

(B) 400,000,000

(C) 3,600,000

(D) 4,000,000,000

$$\frac{1}{x} > -y$$

9. All of the following are solutions to the given equation EXCEPT

(A) $(2, \frac{1}{8})$

(B) $(\frac{1}{8}, 2)$

(C) $(-8, \frac{1}{2})$

(D) $(-\frac{1}{8}, 2)$

10. If $n^2 - p^2 = 40$ and $n + p = 5$, then $n - p =$

(A) 8

(B) 35

(C) 45

(D) 200

11. If the mean of 7, 8, 16, k, and $3k$ equals 16, then $k =$

(A) 11.75

(B) 12.25

(C) 12.75

(D) 13.25

$$2x + 3y = -4$$
$$8x + 7y = 14$$

12. In the linear system of equations shown here, $x + y =$

(A) −2

(B) −1

(C) 1

(D) 2

GO ON TO NEXT PAGE

$$\sqrt{x^2 - 1} - 2 = x$$

13. Which of the following values of x satisfies this equation?

(A) $-\dfrac{1}{4}$

(B) $-\dfrac{3}{4}$

(C) $-\dfrac{5}{4}$

(D) $-\dfrac{7}{4}$

14. If $\sin 5a° = \cos 10b°$, then which of the following equations must be true?

(A) $a + 2b = 9$

(B) $a + 2b = 10$

(C) $a + 2b = 18$

(D) $a + 2b = 36$

$$\left(\dfrac{1}{x}\right)^{-2} = \dfrac{1}{4}$$

15. If this equation is true, then which of the following values could be x?

(A) $\dfrac{1}{16}$

(B) $-\dfrac{1}{2}$

(C) 2

(D) −16

$$2(w+8)-3(w-6)=5(w-4)$$

16. For what value of w is this equation correct?

17. Together, Megan and Andrew sold a total of 63 tickets to their school play. Megan sold 5 fewer than 3 times the number of tickets that Andrew sold. How many tickets did Megan sell?

18. A data set includes 20 values, with a minimum value of 64 and a maximum value of 80. An additional value v is included in the data set and increases the range of the data set by 5. If v is less than the median value of the set, then what does v equal?

$$f(x)=x^3-\frac{1}{2}$$
$$g(x)=4f(x)+1$$

19. Given the two functions defined here, if $g(x)=6x^3-17$, what is the value of x?

$$(x+4)(x+n)=x^2+\frac{px}{32}+\frac{1}{8}$$

20. If this equation is true for all values of x, what is the value of p?

DO NOT TURN THE PAGE UNTIL TOLD TO DO SO **STOP** DO NOT RETURN TO A PREVIOUS TEST

Section 4 — Calculator

TIME: 55 minutes, 38 questions

Directions

For questions 1-30, solve each problem, choose the best answer from the choices provided, and fill in the corresponding bubble on your answer sheet. Refer to the directions before question 31 on how to enter your answers in the grid. You may use any available space in your test booklet for scratch work.

Notes

- The use of a calculator **is permitted.**

- All variables and expressions used represent real numbers unless otherwise indicated.

- Figures provided in this test are drawn to scale unless otherwise indicated.

- All figures lie in a plane unless otherwise indicated.

- Unless otherwise indicated, the domain of a given function f is the set of all real numbers x for which $f(x)$ is a real number.

Reference

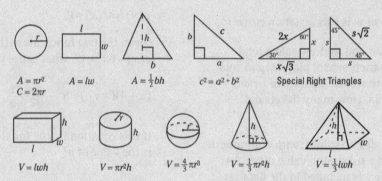

$A = \pi r^2$
$C = 2\pi r$

$A = lw$

$A = \frac{1}{2}bh$

$c^2 = a^2 + b^2$

Special Right Triangles

$V = lwh$

$V = \pi r^2 h$

$V = \frac{4}{3}\pi r^3$

$V = \frac{1}{3}\pi r^2 h$

$V = \frac{1}{3}lwh$

The number of degrees of arc in a circle is 360.

The number of radians of arc in a circle is 2π.

The sum of measures in degrees of the angles of a triangle is 180.

1. If you can exchange one U.S. dollar for 90 Japanese yen, and your hotel room in Tokyo costs 10,800 yen per night, what is the equivalent cost of the room in U.S. dollars?

   (A) $90

   (B) $110

   (C) $120

   (D) $125

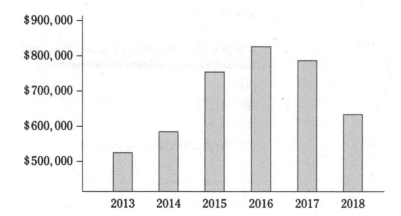

2. This graph shows the amount of money that a company earned during the years 2013 through 2018. In which year did profits increase the most in comparison with the previous year?

   (A) 2014

   (B) 2015

   (C) 2016

   (D) 2017

3. When Sandra uses her riding lawn mower, she can mow 7 square meters in 10 seconds. At this rate, how much time will be required to mow a lawn that is 140 meters long and 36 meters wide?

   (A) 2 hours

   (B) 2 hours and 20 minutes

   (C) 2 hours and 40 minutes

   (D) 3 hours

GO ON TO NEXT PAGE

4. Which of the following graphs is NOT a function?

(A)

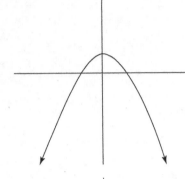

(B)

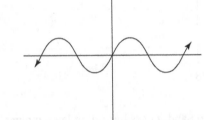

(C)

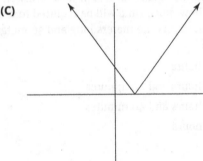

(D)

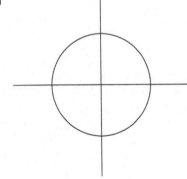

5. If $x = 4$ and $y = -3$, what is the value of $(2xy)^2 - 5y$?

(A) -591

(B) -561

(C) 561

(D) 591

$$x^3 + 32 = -32$$

6. Which of the following is the complete solution set for the equation shown here?

(A) $\{4\}$

(B) $\{-4\}$

(C) $\{-4, 4\}$

(D) The equation has no solution.

x	$f(x)$
2	12
3	30
4	84
5	246

7. This table shows four values for the exponential function $f(x) = 3^x + j$. What is the value of j?

(A) 3

(B) 4

(C) 5

(D) 6

8. Which of the following is the equation of a linear function that has a slope of 3 and passes through the point $(-3, -1)$?

(A) $f(x) = 3x + 2$

(B) $f(x) = 3x + 3$

(C) $f(x) = 3x + 5$

(D) $f(x) = 3x + 8$

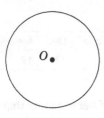

9. The circle shown here centered at O has an area of 8π. What is the circumference of the circle?

    (A) $\sqrt{2}\pi$

    (B) $2\sqrt{2}\pi$

    (C) $3\sqrt{2}\pi$

    (D) $4\sqrt{2}\pi$

10. If $2a + 5 = -3(4b - 5)$, what is the value of $a + 6b$?

    (A) -10

    (B) -5

    (C) 5

    (D) 10

11. A group of friends won d dollars in a raffle. When they collected their winnings, they split the money evenly among four people. Later, they remembered that a fifth person, Justin, had originally helped to purchase the winning ticket. So each of the four people who had split the money gave Justin $40. As a result, each of the five winners received the same amount of money. Which of the following inequalities is correct?

    (A) $500 < d < 750$

    (B) $750 < d < 1{,}000$

    (C) $1{,}000 < d < 1{,}250$

    (D) $1{,}250 < d < 1{,}500$

$$\text{Density} = \frac{\text{Population}}{\text{Land Area}}$$

12. This formula allows you to calculate the density of a city in people per square mile, given its population and its land area in square miles. If Waylonton has a density of 855 people per square mile and a land area of 12.2 square miles, what is its approximate population, to the nearest thousand people?

    (A) 9,000

    (B) 10,000

    (C) 11,000

    (D) 12,000

$$\frac{3x + 3}{x^2 + 3x + 2}$$

13. Which of the following is equivalent to the expression shown here?

    (A) $\dfrac{3}{x + 1}$

    (B) $\dfrac{3}{x + 2}$

    (C) $\dfrac{3}{x^2 + 1}$

    (D) $\dfrac{3}{x^2 + 2}$

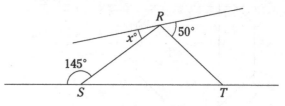

14. In this figure, the length of RS is equivalent to the length of RT. What is the value of x?

    (A) 15

    (B) 20

    (C) 25

    (D) 30

15. One share of stock in EnerTron Industries currently sells for $179. The price of the stock is projected to rise by 6% in each of the next 3 years. If P is the predicted price of the stock, which of the following correctly calculates the expected price of the stock in 3 years?

    (A) $P = 179(0.03)^6$

    (B) $P = 179(1.03)^6$

    (C) $P = 179(0.06)^3$

    (D) $P = 179(1.06)^3$

GO ON TO NEXT PAGE

$$y = 5x^3 + 25x^2 - 80x - 400$$

16. Which of the following is NOT a root of the function shown here?

(A) −5

(B) −4

(C) 4

(D) 5

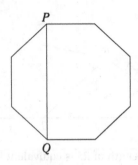

17. This figure shows a regular octagon with sides of length 1. What is the length of PQ?

(A) $1 + \sqrt{2}$

(B) $1 + 2\sqrt{2}$

(C) $1 + \dfrac{\sqrt{2}}{2}$

(D) $1 + \dfrac{\sqrt{2}}{3}$

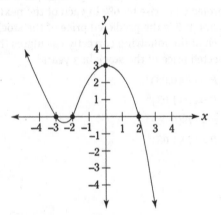

18. Which of the following functions is the equation of this graph?

(A) $y = \dfrac{1}{4}(x+3)(x+2)(x-2)$

(B) $y = \dfrac{1}{4}(x-3)(x+2)(x-2)$

(C) $y = -\dfrac{1}{4}(x+3)(x+2)(x-2)$

(D) $y = -\dfrac{1}{4}(x-3)(x+2)(x-2)$

Tia	Dawn	Rosa	Sylvie	Jane	Livia	Melanie
9	9	8	4	11	12	17

19. This table shows the number of points that 7 college basketball players scored at their most recent game. Which of the following statements is true?

(A) The mean score is 1 greater than the median score.

(B) The mean score is equal to the median score.

(C) The mean score is 1 less than the median score.

(D) The mean score is 2 less than the median score.

$$3x + 4y = 17.3$$
$$-x + 5y = 6.9$$

20. What is the value of x in this system of linear equations?

(A) 2.9

(B) 3.1

(C) 3.4

(D) 3.8

21. Aaron and Zeena are the two finalists in a TV singing competition. Eighty percent of the audience voted using a phone app, and 20% using a computer. If Aaron won 56% of the votes that were registered by app and only 24% of the votes that were registered by computer, then what percentage of the vote did Zeena receive?

(A) 54%

(B) 50.4%

(C) 49.6%

(D) 46%

22. If $i^2 = -1$, then $i^{35} =$

(A) 1

(B) −1

(C) i

(D) −i

$$\frac{2a}{5b} = \frac{1}{4}$$

23. Given this equation, what does $10a - 5b$ equal?

(A) a

(B) $2a$

(C) $3a$

(D) $4a$

24. A rectangular ballroom dance floor has a length that is 4 feet longer than its width. The manager plans to double the area of the floor by adding 10 feet to the length and 9 feet to the width. What is the current area of the dance floor?

(A) 375 square feet

(B) 425 square feet

(C) 475 square feet

(D) 525 square feet

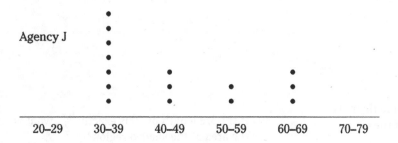

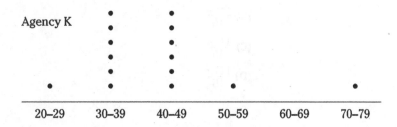

25. Agency J and Agency K both have 15 employees. This histogram shows the ages of these people. Which of the following statements is true about the ages of the people working at these two agencies?

(A) Agency J has a larger range and standard deviation than Agency K.

(B) Agency J has a larger range but a smaller standard deviation than Agency K.

(C) Agency J has a smaller range but a larger standard deviation than Agency K.

(D) Agency J has a smaller range and standard deviation than Agency K.

$$3x^2 - 4y$$

26. Which of the following is equivalent to this expression?

(A) $(\sqrt{3x} + \sqrt{2y})(\sqrt{3x} - \sqrt{2y})$

(B) $(3\sqrt{x} + 2\sqrt{y})(3\sqrt{x} - 2\sqrt{y})$

(C) $(x\sqrt{3} + 2\sqrt{y})(x\sqrt{3} - 2\sqrt{y})$

(D) $(\sqrt{3}x + y\sqrt{2})(\sqrt{3}x - y\sqrt{2})$

GO ON TO NEXT PAGE

$$f(x) = 2x^2 - 20x + 44$$

27. If the vertex of the quadratic equation shown here is at (h, k), then $h + k =$

(A) −1

(B) −3

(C) −5

(D) −7

$$x^2 + 6x + y^2 - 8y = -9$$

28. The equation shown here defines a circle in the xy-plane. What is the radius of the circle?

(A) 3

(B) 4

(C) 5

(D) 6

Questions 29 and 30 refer to the following information.

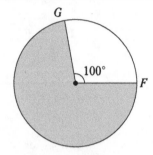

29. If the arc length FG shown in the figure is between 7 and 8, the radius of the circle could be

(A) 3.8 but not 4.2

(B) 4.1 but not 4.5

(C) 4.4 but not 4.8

(D) 4.7 but not 5.1

30. If the arc length of FG is exactly $\frac{5\pi}{2}$, what is the area of the shaded region?

(A) $\frac{117\pi}{8}$

(B) $\frac{117\pi}{10}$

(C) $\frac{117\pi}{15}$

(D) $\frac{117\pi}{16}$

31. If $n = 5$, what is the sum of $n^2 - 10$ and $2n - 3$?

32. If $f(x) = 4x - 1$, what is $f(6) - f(2)$?

$$F + V - E = 2$$

33. This formula works for all solids that have F faces, V vertexes, and E edges. If a solid that has k faces, $k - 8$ vertices, and $k + 10$ edges, then what does k equal?

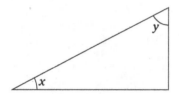

34. In the triangle shown here, $\tan x = \frac{8}{15}$. What is the value of $\cos y$?

35. Ryan read a book in order to write a book report for school. He read the first 70 pages of the book in 105 minutes. Then, he read the next 30% of the book in 126 minutes. Finally, he completed the remaining p pages of the book in m minutes. Assuming that Ryan read all portions of the book at a constant speed, what is the value of $m + p$?

$$ax + 8y = 10$$
$$5x + by = 35$$

36. If the system of linear equations shown here has no solutions, what is the value of ab?

Questions 37 and 38 refer to the following information.

x	$f(x)$
0	1
1	2
2	4
3	0
4	3

$$g(x) = x - \frac{1}{x}$$

37. What is the value of $g(f^{-1}(4))$, given that $f^{-1}(x)$ is the inverse of $f(x)$?

$$f(3) = g(m)$$

38. If the equation shown here is true for some value of $m > 0$, what is the value of m?

Chapter 19

Answers to Practice SAT Math Test 2

Ｈow did your second attempt go? In this chapter, I provide answers and detailed explanations for all the questions in Practice SAT Math Test 2, found in Chapter 18.

No peeking until you're done with that test!

Section 3 — No Calculator

1. **D.** $y = 2x - 1$. The graph has a y-intercept of -1 and a slope of 2, so its equation is $y = 2x - 1$. Therefore, Answer D is correct.

2. **C.** $3\sqrt{2}$. The triangle created by two sides of a square and the diagonal is a $45° - 45° - 90°$ right triangle. When the side of this triangle is x, the length of the diagonal is $x\sqrt{2}$. Thus, when the side of this triangle is 3, the length of the diagonal is $3\sqrt{2}$, so Answer C is correct.

3. **A.** $2xy^2(3y^3 - 5xy + 6x^2)$. All three answers give the greatest common factor of $6xy^5 - 10x^2y^3 + 12x^3y^2$ as $2xy^2$. When you factor out $2xy^2$ from $6xy^5$, the result is $3y^2$. When you factor this value out of $-10x^2y^3$, the result is $-5xy$. And when you factor this value out of $12x^3y^2$, the result is $6x^2$. Thus, the answer is $2xy^2(3y^3 - 5xy + 6x^2)$, so Answer A is correct.

4. **B. May.** According to the table, Victoria deposited $700 into the account every month after the first month, so $t = 700$, and $s < 700$. She had $1,700 in the account in July, so working backward in the table, she had $1,000 in June, and $300 in May. Therefore, she opened the account in May with $300, so Answer B is correct.

5. **A. A y-intercept that is greater than 0 and a minimum value.** The function $f(x) = -3x^2 + 14x + 5$ is a quadratic of the form $f(x) = ax^2 + bx + c$. It has a leading coefficient of $a = -3$, so it's concave down and, therefore, has a maximum value. It has a constant of $c = 5$, so its y-intercept is 5 and, therefore, greater than 0. Thus, Answer A is correct.

6. **C. \$1,400.00.** The cost of the television was \$910, which reflected a 35% discount, so the cost was 65% of the retail price. Thus, let x equal the retail price of the television, so:

$$0.65x = 910$$
$$x = \frac{910}{0.65}$$
$$x = 1,400$$

Therefore, the Answer is C.

7. **B. 5 to 1.** According to the graph, the population of China will be a little less than 1.5 billion people, and that of Indonesia a little less than 0.3 billion people. To find the ratio, divide 1.5 by 0.3 and then convert the result to a fraction:

$$\frac{1.5}{0.3} = 5 = \frac{5}{1}$$

Thus, the ratio is approximately 5 to 1, so Answer B is correct.

8. **C. 3,600,000,000.** According to the graph, the population of China will be 1,500,000,000, that of India will be 1,400,000,000, that of the United States will be 400,000,000, and that of Pakistan will be 300,000,000. Thus, the sum of these values is:

$$1,500,000,000 + 1,400,000,000 + 400,000,000 + 300,000,000 = 3,600.000.000$$

Thus, even allowing for reasonable errors in approximation, Answer C is correct.

9. **D. $(-\frac{1}{8}, 2)$.** If both x and y are positive, the left side of the inequality $\frac{1}{x} > -y$ will be positive and the right side negative. This rules out Answers A and B. To decide between Answers C and D, plug both pairs of values into the inequality and simplify:

$$-\frac{1}{8} > -\frac{1}{2} \qquad \qquad \frac{1}{-\frac{1}{8}} > -2 \quad \text{FALSE!}$$
$$-8 > -2$$

Therefore, Answer D is correct.

10. **A. 8.** You can factor $n^2 - p^2$ as follows:

$$n^2 - p^2 = (n + p)(n - p)$$

Now, plug in 40 for $n^2 - p^2$ and 5 for $n + p$:

$$40 = 5(n - p)$$

Next, divide both sides by 8:

$$8 = n - p$$

Therefore, Answer A is correct.

11. **B. 12.25.** If the mean of 5 numbers equals 16, then the sum of those 5 numbers equals $16 \times 5 = 80$. Thus, you can make and solve the following equation:

$$7 + 8 + 16 + k + 3k = 80$$
$$31 + 4k = 80$$
$$4k = 49$$
$$k = 12.25$$

Therefore, Answer B is correct.

12. **C. 1.** To begin, multiply the first equation in the system by -4:

$$-8x - 12y = 16$$
$$8x + 7y = 14$$

Next, add the two equations together to form a single equation in one variable, and solve:

$$-5y = 30$$
$$y = -6$$

Now, plug in -6 for y into the first equation:

$$2x + 3(-6) = -4$$
$$2x - 18 = -4$$
$$2x = 14$$
$$x = 7$$

Thus, $x + y = 7 + (-6) = 1$, so Answer C is correct.

13. **C.** $-\dfrac{5}{4}$. To begin, add 2 to both sides of the equation:

$$\sqrt{x^2 - 1} - 2 = x$$
$$\sqrt{x^2 - 1} = x + 2$$

Next, square both sides and simplify:

$$(\sqrt{x^2 - 1})^2 = (x + 2)^2$$
$$x^2 - 1 = (x + 2)(x + 2)$$
$$x^2 - 1 = x^2 + 4x + 4$$

Now, subtract x^2 from both sides of the equation and solve for x:

$$-1 = 4x + 4$$
$$-5 = 4x$$
$$-\frac{5}{4} = x$$

Therefore, Answer C is correct.

14. **C.** $a + 2b = 18$. Begin by recalling the trig identity $\sin x° = \cos(90 - x)°$. Because $\sin 5a° = \cos 10b°$, you can make the following equation:

$$5a + 10b = 90$$

Now, divide both sides by 5:

$$a + 2b = 18$$

Therefore, Answer C is correct.

15. **B.** $-\dfrac{1}{2}$. To begin, change the negative exponent to a positive one by using the reciprocal of the base:

$$(\frac{1}{x})^{-2} = \frac{1}{4}$$
$$x^2 = \frac{1}{4}$$

Now, take the square root of both sides:

$$\sqrt{x^2} = \pm\sqrt{\frac{1}{4}}$$
$$x = \pm\frac{\sqrt{1}}{\sqrt{4}}$$
$$x = \pm\frac{1}{2}$$

Therefore, $-\dfrac{1}{2}$ is one of the two possible values of x, so Answer B is correct.

16. **9.** Begin simplifying the equation by distributing wherever possible and then combining like terms:

$$2(w+8) - 3(w-6) = 5(w-4)$$
$$2w + 16 - 3w + 18 = 5w - 20$$
$$-w + 34 = 5w - 20$$

Next, isolate w and solve:

$$34 = 6w - 20$$
$$54 = 6w$$
$$9 = w$$

Therefore, the correct answer is 9.

17. **46.** To begin, let a equal the number of tickets that Andrew sold. Megan sold 5 fewer than 3 times the number of tickets that Andrew sold, so the number of tickets Megan sold is $3a - 5$. Together, Megan and Andrew sold 63 tickets, so you can make the following equation:

$$a + 3a - 5 = 63$$

Simplify and solve for a:

$$4a - 5 = 63$$
$$4a = 68$$
$$a = 17$$

Thus, Andrew sold 17 tickets, so Megan sold $63 - 17 = 47$ tickets, and so the answer is 46.

18. **59.** A data set that has a minimum value of 64 and a maximum value of 80 has a range of $80 - 64 = 16$. Thus, when the additional value v is included in the data set, the range of the set increases to $16 + 5 = 21$. The value of v is less than the median of the set, so it's less than 80. Thus, v is 5 less than the least value in the data set, so $v = 64 - 5 = 59$. This increases the range of the set to $80 - 59 = 21\ 80 - 59 = 29$, so the answer is 59.

19. **2.** To begin, find $g(x)$ by plugging in $x^3 - \dfrac{1}{2}$ for $f(x)$:

$$g(x) = 4(x^3 - \frac{1}{2}) + 1 = 4x^3 - 2 + 1 = 4x^3 - 1$$

Additionally, the problem gives you the following equation:

$$g(x) = 6x^3 - 17$$

Thus, you can set these two values of $g(x)$ equal and solve for x:

$$6x^3 - 17 = 4x^3 - 1$$
$$2x^3 = 16$$
$$x^3 = 8$$
$$x = 2$$

Therefore, the answer is 2.

20. 129. To begin, FOIL the left side of the equation:

$$(x+4)(x+n) = x^2 + \frac{px}{32} + \frac{1}{8}$$
$$x^2 + nx + 4x + 4n = x^2 + \frac{px}{32} + \frac{1}{8}$$

This equation is true for all values of x, so the constant values $4n$ and $\frac{1}{8}$ are also equal, and so

$$4n = \frac{1}{8}$$
$$n = \frac{1}{32}$$

Thus, you can substitute $\frac{1}{32}$ for n into the equation:

$$x^2 + \frac{x}{32} + 4x + \frac{4}{32} = x^2 + \frac{px}{32} + \frac{1}{8}$$
$$x^2 + \frac{x}{32} + 4x + \frac{1}{8} = x^2 + \frac{px}{32} + \frac{1}{8}$$

Subtract out $x^2 + \frac{1}{8}$ on both sides of the equation:

$$\frac{x}{32} + 4x = \frac{px}{32}$$

Divide out an x:

$$\frac{1}{32} + 4 = \frac{p}{32}$$

Multiply both sides by 32 to remove the fractions, and solve for p:

$$\frac{1}{32} + 4 = \frac{p}{32}$$
$$1 + 128 = p$$
$$129 = p$$

Therefore, the answer is 129.

Section 4 — Calculator

1. **C. $120.** To convert 10,800 yen to U.S. dollars, multiply by the conversion factor $\dfrac{\$1}{90\text{ yen}}$:

$$10,800 \text{ yen} \times \frac{\$1}{90 \text{ yen}} = \frac{\$10,800}{90} = \$120$$

So Answer C is correct.

2. **B. 2015.** The greatest increase in profits between consecutive years occurred from 2014 to 2015, an increase of approximately $200,000. Thus, Answer B is correct.

3. **A. 2 hours.** In 10 seconds, Sandra can mow 7 square meters. Thus, in 1 minute (60 seconds), she can mow $7 \times 6 = 42$ meters. The area of the lawn is $140 \times 36 = 5,040$ square meters. To calculate the number of minutes required to mow this lawn, divide 5,040 by 42:

$$5,040 \div 42 = 120$$

Therefore, to mow the lawn requires 120 minutes, which equals 2 hours, so Answer A is correct.

4. **D.**

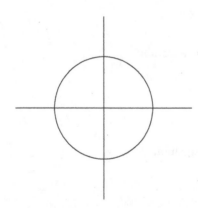

A graph is a function if and only if it has no more than one output (y) value for every input (x) value. Another way to state this is that a graph is a function if and only if you cannot pass a vertical line through more than one point on that graph. Of the four graphs, only the circle in Answer D is not a function, so Answer D is correct.

5. **D. 591.** Plug in 4 for x and -3 for y into $(2xy)^2 - 5y$ and simplify:

$$[2(4)(-3)]^2 - 5(-3) = (-24)^2 + 15 = 576 + 15 = 591$$

Therefore, Answer D is correct.

6. **B. {−4}.** Solve the equation by subtracting 32 from both sides, and then taking the cube root of the result:

$$\begin{aligned} x^3 + 32 &= -32 \\ x^3 &= -64 \\ \sqrt[3]{x^3} &= \sqrt[3]{-64} \\ x &= -4 \end{aligned}$$

This is the only solution to the equation, so Answer B is correct.

7. A. 3. To find the value of j, plug in any coordinate pair in the table into the function. For example, here's the result when you plug in the coordinate pair (2,12):

$$f(x) = 3^x + j$$
$$12 = 3^2 + j$$
$$12 = 9 + j$$
$$3 = j$$

Therefore, Answer A is correct.

8. D. $f(x) = 3x + 8$. A linear function with a slope of 3 has the following equation:

$$y = 3x + b$$

To solve for b, substitute -3 for x and -1 for y into the equation:

$$-1 = 3(-3) + b$$
$$-1 = -9 + b$$
$$8 = b$$

Thus, the equation for the line is $y = 3x + 8$. In function form, this equation is $f(x) = 3x + b$, so Answer D is correct.

9. D. $4\sqrt{2}\pi$. To begin, use the area formula for a circle to find the radius:

$$A = \pi r^2$$
$$8\pi = \pi r^2$$
$$8 = r^2$$
$$\sqrt{8} = r$$

Now, plug this value into the circumference formula for a circle:

$$C = 2\pi r = 2\sqrt{8}\pi$$

To complete the problem, factor and simplify the radical:

$$2\sqrt{4}\sqrt{2}\pi = 4\sqrt{2}\pi$$

Therefore, Answer D is correct.

10. C. 5. To begin, distribute the right side of the equation and simplify:

$$2a + 5 = -3(4b - 5)$$
$$2a + 5 = -12b + 15$$
$$2a = -12b + 10$$

Now, add $12b$ to both sides:

$$2a + 12b = 10$$

Now, to make the left side of the equation equal to $a + 6b$, divide both sides of the equation by 2:

$$a + 6b = 5$$

Therefore, Answer C is correct.

11. **B. $750 < d < 1{,}000$.** The group of friends split d dollars four ways, so each person received $\dfrac{d}{4}$ dollars. Then, each person gave Justin \$40, so each of the five people (including Justin) had $\dfrac{d}{4} - 40$ dollars. Justin received \$40 from each person, so he had \$160. Thus, you can make and solve the following equation:

$$\frac{d}{4} - 40 = 160$$
$$\frac{d}{4} = 200$$
$$d = 800$$

Therefore, $750 < d < 1{,}000$, so Answer B is correct.

12. **B. 10,000.** Plug in 855 for density and 12.2 for land area into the formula $\text{Density} = \dfrac{\text{Population}}{\text{Land Area}}$:

$$855 = \frac{\text{Population}}{12.2}$$

To solve for population, multiply both sides by 12.2:

$$855 \times 12.2 = \frac{\text{Population}}{12.2} \times 12.2$$
$$10{,}431 = \text{Population}$$

Rounded to the nearest value, 10,431 becomes 10,000, so Answer B is correct.

13. **B. $\dfrac{3}{x+2}$.** To begin, factor out a GCF of 3 in the numerator, and factor the quadratic in the denominator into two binomials:

$$\frac{3x+3}{x^2+3x+2} = \frac{3(x+1)}{(x+1)(x+2)}$$

Now, cancel a factor of $x+1$ in the numerator and denominator:

$$= \frac{3}{x+2}$$

Therefore, Answer B is correct.

14. **B. 20.** The 145° angle and angle RST are a linear pair, so the measure of angle RST is 35°. The length of RS equals that of RT, so triangle RST is isosceles. Thus, the two lower angles of the triangle are equal, so angle STR also measures 35°. The sum of three angles in a triangle is 180°, so angle TRS measures $180° - 35° - 35° = 110°$. So, you can solve the following equation to find x:

$$x° + 110° + 50° = 180°$$
$$x° + 160° = 180°$$
$$x° = 20°$$

Therefore, Answer B is correct.

15. **D. $P = 179(1.06)^3$.** The price of the stock is expected to rise by 6% per year, so the base of the exponent must be 1.06. Thus, Answer D is correct.

16. D. 5. Begin by setting y to 0, and divide both sides of the equation by 5:

$$0 = 5x^3 + 25x^2 - 80x - 400$$
$$0 = x^3 + 5x^2 - 16x - 80$$

Factor the cubic polynomial on the right side of the equation by grouping:

$$0 = x^2(x+5) - 16(x+5)$$
$$0 = (x^2 - 16)(x+5)$$

Now, factor the expression $x^2 - 16$ as a difference of squares:

$$0 = (x+4)(x-4)(x+5)$$

Now, split this equation into three separate equations and solve for x:

$$x+4 = 0 \qquad x-4 = 0 \qquad x+5 = 0$$
$$x = -4 \qquad x = 4 \qquad x = -5$$

Thus, -4, 4, and -5 are all roots of the original function, so Answer D is correct.

17. A. $1 + \sqrt{2}$. Begin by drawing two additional horizontal lines in the figure:

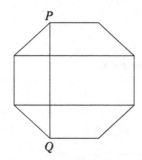

Now, PQ is divided into three smaller segments. The length of the middle segment equals the side of the octagon, which is 1.

Each of the two smaller segments is a leg of a $45° - 45° - 90°$ right triangle whose hypotenuse equals 1 (the side of the octagon). The proportions of the three sides of this triangle are $1 : 1 : \sqrt{2}$, so letting x equal the leg of this triangle, you can make the following proportional equation:

$$\frac{x}{1} = \frac{1}{\sqrt{2}}$$
$$x = \frac{1}{\sqrt{2}}$$

You can rationalize this denominator as follows:

$$x = \frac{1}{\sqrt{2}} \times \frac{\sqrt{2}}{\sqrt{2}} = \frac{\sqrt{2}}{2}$$

Thus, the length of PQ is 1 (the middle segment) plus two segments of length $\frac{\sqrt{2}}{2}$, so it equals:

$$1 + \frac{\sqrt{2}}{2} + \frac{\sqrt{2}}{2} = 1 + \sqrt{2}$$

Therefore, Answer A is correct.

18. **C.** $y = -\frac{1}{4}(x+3)(x+2)(x-2)$. The end behavior of the graph is negative when x increases, so the function is negative, which rules out Answers A and B. The graph has a root at $x = -3$, so it includes a factor of $(x+3)$. Therefore, Answer C is correct.

19. **A. The mean score is 1 greater than the median score.** To find the mean score, add up the scores and divide by the number of players, which is 7:

$$\text{Mean} = \frac{9+9+8+4+11+12+17}{7} = \frac{70}{7} = 10$$

To find the median score, arrange the 7 scores from least to greatest, and select the middle value:

4, 8, 9, 9, 11, 12, 17

So the mean score is 10 and the median score is 9; therefore, Answer A is correct.

20. **B. 3.1.** To begin, multiply the second equation by 3:

$$3x + 4y = 17.3$$
$$-3x + 15y = 20.7$$

Next, add the two equations together and solve for y:

$$19y = 38$$
$$y = 2$$

Plug in 2 for y into the second equation and solve for x:

$$-x + 5(2) = 6.9$$
$$-x + 10 = 6.9$$
$$-x = -3.1$$
$$x = 3.1$$

Therefore, Answer B is correct.

21. **B. 50.4%.** To calculate the number of votes that Aaron received by phone app, multiply as follows:

$$0.80 \times 0.56 = 0.448 = 44.8\%$$

To calculate the number of votes he received by computer, multiply as follows:

$$0.20 \times 0.24 = 0.848 = 4.8\%$$

Thus, Aaron received $44.8\% + 4.8\% = 49.6\%$ of the vote. Therefore, Zeena received $100\% - 49.6\% = 50.4\%$, so Answer B is correct.

22. **D.** $-i$. To begin, factor i^{35} as follows:

$$i^{35} = i^{32}i^3$$

The value of i raised to any power that's a multiple of 4 is 1, so:

$$i^{32}i^3 = i^3$$

Now, factor i^3 as follows:

$$i^3 = i^2 i$$

The problem states that $i^2 = -1$, so:

$$i^2 i = (-1)i = -i$$

Therefore, $i^{35} = -i$, so Answer D is correct.

23. **B. $2a$.** Cross-multiply the equation to get rid of the fractions:

$$\frac{2a}{5b} = \frac{1}{4}$$
$$8a = 5b$$

Next, subtract $5b$ over to the left side of the equation:

$$8a - 5b = 0$$

To change the left side of the equation to $10a - 5b$, add $2a$ to both sides of the equation:

$$10a - 5b = 2a$$

Therefore, Answer B is correct.

24. **D. 525 square feet.** Let w equal the current width of the dance floor. So $w + 4$ equals the length, and the current area is:

$$\text{Current Area} = w(w + 4)$$

If the width is increased by 9 feet and the length by 10 feet, the projected area of the dance floor will be:

$$\text{Projected Area} = (w + 9)(w + 14)$$

The projected area will be twice the size of the current area, so:

$$2w(w + 4) = (w + 9)(w + 14)$$

To solve this equation, begin by simplifying and factor the result:

$$2w^2 + 8w = w^2 + 23w + 126$$
$$w^2 - 15w - 126 = 0$$
$$(w + 6)(w - 21) = 0$$

To finish solving, set each factor equal to 0 and solve:

$$w + 6 = 0 \qquad w - 21 = 0$$
$$w = -6 \qquad w = 21$$

The current width of the floor must be a positive number, so it's 21 feet. Thus, the current length is $21 + 4 = 25$ feet, and the current area is $21 \times 25 = 525$ square feet. Therefore, Answer D is correct.

25. **C. Agency J has a smaller range but a larger standard deviation than Agency K.** The histogram for Agency J shows a range of ages from 30 to 69, so this range is at most 39. The histogram for Agency K shows a range of ages from 29 to 70, so this range is at least 41. Thus, Agency J has a smaller range than Agency K, which rules out Answers A and B.

The histogram for Agency K shows a much greater concentration of employees toward the center of the age range than Agency J. Thus, Agency J has a larger standard deviation than Agency K. This rules out Answer D, so Answer C is correct.

26. **C. $(x\sqrt{3}+2\sqrt{y})(x\sqrt{3}-2\sqrt{y})$.** The four answers resemble the following difference of squares formula:

$$a^2 - b^2 = (a+b)(a-b)$$

To use this formula to find an equivalent expression to $3x^2 - 4y$, first find the square root of $3x^2$ and $4y$:

$$\sqrt{3x^2} = x\sqrt{3} \qquad\qquad \sqrt{4y} = 2\sqrt{y}$$

Plug in these two values as a and b into the formula:

$$3x^2 - 4y = (x\sqrt{3}+2\sqrt{y})(x\sqrt{3}-2\sqrt{y})$$

Therefore, Answer C is correct.

27. **A. −1.** To find the vertex of $f(x) = 2x^2 - 20x + 44$, begin by finding the axis of symmetry:

$$x = -\frac{b}{2a} = -\frac{-20}{2(2)} = \frac{20}{4} = 5$$

Thus, the x-coordinate of the vertex is $h = 5$. To find the corresponding y-coordinate, plug in this value for x into the function and simplify:

$$f(x) = 2(5)^2 - 20(5) + 44 = 50 - 100 + 44 = -6$$

Thus, $k = -6$, so:

$$h + k = 5 + (-6) = -1$$

Therefore, Answer A is correct.

28. **B. 4.** To find the radius, you need to put the circle in center-radius form $(x-h)^2 + (y-k)^2 = r^2$ by completing the square for both x and y. To complete the square for x, add 9 to both sides and then factor the first three terms as follows:

$$x^2 + 6x + 9 + y^2 - 8y = -9 + 9$$
$$(x+3)^2 + y^2 - 8y = 0$$

To complete the square for y, add 16 to both sides and then factor the last three terms:

$$(x+3)^2 + y^2 - 8y + 16 = 16$$
$$(x+3)^2 + (y-4)^2 = 16$$

Thus, $r^2 = 16$, so $r = 4$; therefore, Answer B is correct.

29. **C. 4.4 but not 4.8.** To begin, change $100°$ to radians:

$$100° \times \frac{\pi}{180°} = \frac{5\pi}{9} \approx 1.74$$

Plug in this value for radians and 7 for arc length into the arc length formula Arc Length = Radius × Radians, solving for *radius*:

$$7 = \text{Radius} \times 1.74$$
$$4.02 \approx \text{Radius}$$

Now, repeat this process, using 8 for arc length:

$$8 = \text{Radius} \times 1.74$$
$$4.59 \approx \text{Radius}$$

Thus, the approximate range for the radius of the circle is:

$$4.02 \leq \text{Radius} \leq 4.59$$

Therefore, Answer C is correct.

30. A. $\dfrac{117\pi}{8}$. Plug the value $\dfrac{5\pi}{9}$ for the angle in radians and 2.5π for arc length into the arc length formula, and solve for the radius:

$$\frac{5\pi}{2} = \text{Radius} \times \frac{5\pi}{9}$$
$$\frac{9}{5\pi} \times \frac{5\pi}{2} = \text{Radius}$$
$$\frac{9}{2} = \text{Radius}$$

Now plug this value into the area formula for a circle:

$$\text{Area} = \pi r^2 = \pi \left(\frac{9}{2}\right)^2 = \frac{81\pi}{4}$$

This is the area of the complete circle. The shaded area includes 260° of the total 360°, which makes it $\dfrac{13}{18}$ of the circle, so multiply the area $\dfrac{81\pi}{4}$ by $\dfrac{13}{18}$:

$$\frac{81\pi}{4} \times \frac{13}{18} = \frac{9\pi}{4} \times \frac{13}{2} = \frac{117\pi}{8}$$

Therefore, Answer A is correct.

31. 22. Make a single expression that is the sum of $n^2 - 10$ and $2n - 3$:

$$n^2 - 10 + 2n - 3$$

Substitute 5 for n into this expression and evaluate:

$$= 5^2 - 10 + 2(5) - 3 = 25 - 10 + 10 - 3 = 22$$

Therefore, the answer is 22.

32. 16. Begin by calculating $f(6)$ and $f(2)$:

$$f(6) = 4(6) - 1 = 23 \quad f(2) = 4(2) - 1 = 7$$

Now, calculate $f(6) - f(2)$:

$$f(6) - f(2) = 23 - 7 = 16$$

Therefore, the answer is 16.

33. 20. Plug in k for F, $k-8$ for V, and $k+10$ for E into the formula $F+V-E=2$:

$$F+V-E=2$$
$$k+(k-8)-(k+10)=2$$
$$k+k-8-k-10=2$$
$$k-18=2$$
$$k=20$$

Therefore, the answer is 20.

34. $\dfrac{8}{17}$ **or .470 or .471.** The equation $\tan x = \dfrac{8}{15}$ means that the short and long legs of the triangle are in an 8-to-15 ratio. Thus, the triangle is an $8-15-17$ triangle. You can also find this out by using the Pythagorean Theorem $a^2+b^2=c^2$:

$$8^2+15^2=c^2$$
$$64+225=c^2$$
$$289=c^2$$
$$17=c$$

Thus, $\cos y = \dfrac{8}{17} \approx 0.4705$, which you can grid into the chart as either .470 or .471.

35. 315. Ryan read the 70 pages of the book in 105 minutes, so he read at a rate of $105 \div 70 = 1.5$ minutes per page. He read the next 30% of the book in 126 minutes, so you can set up a proportion to find out what percentage of the book he read first:

$$\frac{x}{105}=\frac{30}{126}$$
$$126x=3150$$
$$x=25$$

Thus, 70 pages comprised 25% of the book, so the entire book consisted of 280 pages. The remaining p pages of the book comprised the remaining 45% of the book, so:

$$p=280\times0.45=126$$

He took 1.5 minutes to read each of these pages, so:

$$m=126\times1.5=189$$

Thus, $m+p=126+189=315$, so the answer is 315.

36. 40. When a system of linear equations has no solutions, the values of its x and y coordinates are proportional in both equations. Thus, in this system of equations, the following equation is true:

$$\frac{a}{5}=\frac{8}{b}$$

Simplify this equation by cross-multiplying:

$$ab=40$$

Therefore, the answer is 40.

37. $\dfrac{3}{2}$ **or 1.5.** $f^{-1}(x)$ is the inverse of $f(x)$, so it accepts an input $f(x)$ and outputs x. Thus, $f^{-1}(4)=2$. Next, substitute 2 for $f^{-1}(4)$ into the expression $g(f^{-1}(4))$:

$$g(f^{-1}(4))=g(2)$$

Now, use $g(x) = x - \dfrac{1}{x}$ to find $g(2)$:

$$g(2) = 2 - \dfrac{1}{2} = \dfrac{3}{2}$$

Therefore, the answer is $\dfrac{3}{2}$ or 1.5.

38. **1.** To begin, use the table to evaluate $f(3)$:

$$f(3) = 0$$

Thus, you're looking for a positive value of m such that

$$g(m) = 0$$

Now, use $g(x) = x - \dfrac{1}{x}$ to evaluate $g(m)$:

$$g(m) = m - \dfrac{1}{m}$$

Set this expression equal to 0 and solve for m:

$$m - \dfrac{1}{m} = 0$$
$$m = \dfrac{1}{m}$$
$$m^2 = 1$$
$$m = \pm\sqrt{1}$$
$$m = \pm 1$$

Therefore, the answer is 1.

6

The Part of Tens

Chapter 20

Ten Things You Will *Almost Definitely* See on Your SAT Math Test

Although the math you need to do well on the SAT may look endless, a lot of SAT questions draw from a relatively small pool of math topics.

In this chapter, you discover what I believe are the ten most common SAT math topics. You're just about dead certain to encounter most or all of them when you take the SAT. I also tell you where in this book you can find more information on each topic.

Linear Functions

A *linear function* has the form $y = mx + b$, and shows up as the graph of a straight line on the xy-plane.

Linear functions are probably the most common topic on the SAT. To answer these questions, you should know how to represent a single linear function in four distinct ways:

» In words

» In a table

» As a graph on the xy-plane

» As an equation

I also recommend knowing how to find the slope of a linear function using the equation $m = \frac{y_2 - y_1}{x_2 - x_1}$.

It's also a good idea to know how to convert a linear function from standard form $ax + by = c$ into the more useful slope-intercept form shown here.

In Chapter 5, I pull together all these formulas and concepts.

Systems of Linear Equations

A *system of linear equations* contains at least two equations in at least two variables. For example:

$$3x + 2y = 7$$
$$-5x + 4y = 3$$

Systems of linear equations are second in frequency only to linear functions as a topic of SAT questions.

Some SAT questions give you a system like this one and ask you to solve it for x, y, or both. In other cases, the question may present you with a word problem and ask you either to solve it using a system of equations or to identify the correct system. Additionally, some SAT questions require you to work with systems that have either no solution or infinitely many solutions.

In Chapter 6, you get up to speed on these key issues involving systems of linear equations.

Quadratic Functions

A *quadratic function* has the form $y = ax^2 + bx + c$, and on the xy-plane takes the shape of a parabola.

Questions about quadratic functions are almost as common as those about systems of linear equations. On the SAT, you'll need to know how to find the y-intercept, concavity (up or down), axis of symmetry, and coordinates of the vertex. It's also helpful to know how to work with a quadratic function in vertex form $y = a(x - h)^2 + k$, as well as how to find the roots of a quadratic function by solving its related quadratic equation $0 = ax^2 + bx + c$.

See Chapter 12 for a complete look at what you need to know about quadratic functions for the SAT.

Percentages

Although you probably first learned to work with percentages in fourth or fifth grade, the SAT includes a variety of questions that lean heavily on them.

In some cases, you may need to use algebra to discover an unnamed percentage (x%). In others, an SAT question may ask you to identify percent increase or percent decrease. Additionally, questions that involve exponential growth or decay functions often test your ability to represent percent increase or decrease as a decimal.

To beef up on your knowledge of percentages, check out Chapter 7.

Ratio/Proportion Problems

A *ratio* is relation between two numbers, similar to a fraction. For example, a ratio of 1 to 3 (or 1:3) implies that one number is $\frac{1}{3}$ of another number.

A ratio allows you to set up a *proportion*, or *proportional equation*, to solve a problem. For example, if the proportion of nurses to support staff at a hospital is 1 to 3, then you can set up the following proportion:

$$\frac{\text{Nurses}}{\text{Support staff}} = \frac{1}{3}$$

Proportions are useful for answering a wide variety of SAT questions. Chapter 7 gives you a look at how to work with ratios and proportions on the SAT.

The Mean and the Median

The *mean* is the average value of a set of numbers. You can calculate it using the following formula:

$$\text{Mean} = \frac{\text{Sum of values}}{\text{Number of values}}$$

In contrast, the *median* is the middle value of a set of numbers (or the *mean* of the two middle values). To find it, you arrange the values in order from smallest to largest, and count your way to the middle.

The mean and the median are two of the most basic concepts in statistics, and you're very likely to need them when you take the SAT.

To discover what you need to know about the mean and the median, plus other important concepts in statistics and probability, see Chapter 8.

The Trig Identity

Don't ask me why, but the makers of the SAT *love* this trig identity:

$$\sin x° = \cos(90 - x)°$$

They love it so much that it appears on almost every test in some form or another. That's why I always tell my students that if they know just one fact about trigonometry, they should be sure to make it this one.

I discuss this identity and a whole lot more about trigonometry in Chapter 14.

The Quadratic Formula

The *quadratic formula* allows you to solve a quadratic equation of the form $0 = ax^2 + bx + c$ for x. Here it is in all its glory:

$$x = \frac{-b \pm \sqrt{b^2 - 4ac}}{2a}$$

In the good old days, way back in 2014, the SAT didn't require you to know the quadratic formula. Now it's part of the test, and you'll probably need it to answer at least one question.

For more on the quadratic formula, plus other important information to help you solve algebra equations, flip to Chapter 4.

The Imaginary Number *i*

The imaginary number i equals $\sqrt{-1}$. It is a number that, strangely, doesn't appear on the real number line.

Complex numbers are numbers of the form $a + bi$. On the SAT, you may need to add, subtract, multiply, or divide complex numbers. You may also need to know how to raise i to the power of an integer.

Check out Chapter 15 for more on imaginary and complex numbers.

Circles on the Graph

Here's the formula for a circle of radius r centered at the origin:

$$x^2 + y^2 = r^2$$

And here's the more general formula for a circle of radius r that's centered at any point (h, k) on the xy-plane:

$$(x - h)^2 + (y - k)^2 = r^2$$

For more on graphing circles, see Chapter 15.

Chapter **21**

Ten Tips to Improve Your SAT Math Score

What are the best uses of your precious time while studying for the math sections of the SAT? In this chapter, I list what I consider to be the ten most important practices you can implement to garner the skills you need most.

Study Diligently in Your Math Classes

Getting a good score on the SAT isn't a trick. The SAT does a very good job of testing what you really know about math. No amount of "guessing (C)" is going to get you a good score if your math skills are nonexistent.

Fortunately, your current math class is the one place that you can go where, for no money out of your pocket, you can build these skills. Even if the curriculum in your current class is beyond what's tested on the SAT — for example, if you're studying calculus or another more advanced Pre-Calculus topic— staying current is going to help you. (And blowing it off isn't going to help your SAT score, your grades, or your readiness for college.)

Look at it this way: If you're looking to get into college, you need both a good SAT score *and* good grades. Studying hard in your math classes will pay off on both fronts.

Get Good at Doing Basic Calculations in Your Head

There's a big difference between knowing something and knowing it cold. Chances are, you know what $2+2$ equals without even thinking about it. But do you get thrown by $-8-(-6)$? Do you feel confident about your ability to factor $6x^2 - 4xy$ without stress?

In Chapter 2, I give you a refresher on a set of Pre-Algebra skills that are important on the SAT. Then in Chapters 3 and 4, I run thorough a whole year's worth of Algebra 1 skills that you'll need to do well. Get comfortable with these moves and you'll significantly improve your SAT score.

Get Good at Using Your Calculator

Calculators — including graphing calculators and scientific calculators — are allowed on the SAT Calculator Section, providing you with a great opportunity to save time and avoid mistakes. Although you don't want to feel forced to use your calculator for simple calculations that you could do more quickly in your head, I encourage you to use it whenever it may be really helpful.

In Chapter 2, I provide a list of the calculator skills I believe are most helpful on the SAT. These include working with fractions, calculating powers and square roots, solving and graphing equations, finding the roots of equations, and generating input-output tables.

Study SAT-Specific Math Skills

The SAT tests a relatively limited set of skills from Pre-Algebra, Algebra I, Algebra II, and a tiny bit of Geometry. These skills include linear and quadratic equations, functions, polynomials, statistics and probability, and a basic understanding of more advanced topics like complex numbers and trigonometry.

While this is admittedly a chunk of information, a lot of what you study in high school is left out. For example, you don't need to know much about graphing trigonometric functions, to do well on the SAT. In geometry, you need to know how to find the angles of a triangle, but you can safely avoid just about everything you know about doing geometric proofs. And as for the tough stuff you struggled to learn about rational functions, radicals, logarithms, the complex plane, or polar coordinates (what was DeMoivre's Theorem again?), you won't go anywhere near it on the SAT.

In this book, I cover the basic toolset that you need to answer most SAT questions. In Chapters 3 through 15, I break down these skills and subskills. Focusing on strengthening these SAT-specific math skills is time well spent as you prepare for the test.

Get Comfortable Turning Words into Numbers

In a sense, word problems are a very specific type of reading comprehension question. You need to become good at the careful type of reading that allows you to turn information in a word problem into numbers, symbols, and equations. In many cases, you may find that after turning words into numbers, the rest of a word problem is a lot easier than it looks, and you can solve it easily in your head or with a calculator.

Throughout the book, I include a variety of word problems in just about every chapter. Don't skip over them! Instead, take time to read through them so that when you take your SAT, you'll feel comfortable bridging the divide between words and numbers.

Take Timed Practice Tests

No matter how good your math skills are, you should practice for the SAT with timed tests. Time pressure adds a dimension to testing that isn't normally present when you're studying. It also forces you to make trade-offs, such as skipping over a problem that looks difficult or time-consuming.

This book contains hundreds of questions to practice on without the clock running. But it also includes two full timed practice tests, plus an additional online practice test, with a total of 174 questions. I highly recommend saving these questions for when you're ready to practice under the timed conditions stated at the top of each test. You can also visit sat.collegeboard.com to get a free, official SAT practice test from the College Board.

When Taking Practice Tests, Budget Your Time to Maximize Your Score

Both the No Calculator and Calculator math sections of the SAT have two parts, the first with multiple-choice questions and the second with grid-in questions. Each of these four parts present questions in order of difficulty from easy to medium to hard.

Using this information to your advantage will help you get the highest possible score. Remember, you don't have to answer every question on the test to get a good score. In fact, answering a few more than half of the questions correctly will get you a score of 500, which might be enough to get you into the college you're aiming for.

So budget your time so that you can focus on correctly answering as many easy and medium questions as you can. Then, work on only the hard questions that you think you can solve. As for the rest, guess and move on.

Study from Your Timed Practice Tests

After you take a practice test, go over the answers you got wrong and find out why. Use the test to fill in gaps in your knowledge that may be useful on the next test.

Remember, any math skill that showed up on one question is likely to show up on a later SAT — possibly yours. Furthermore, as you spend time examining SAT questions in depth, you'll begin to get a sense of how they're put together, giving you an advantage in answering questions you haven't seen.

Retake Your Timed Practice Tests

In studying from the practice tests you've already taken, did you really absorb new material? One way to find out is to take the test again. My advice is to wait a few weeks so that you forget the specifics of each question. In the meantime, take a few more practice tests and study those, too. Then go back and retake a test. Your score will almost certainly be better than your score the first time you took it. But take a good, hard look at the questions you miss the second time around to be sure you know what you need to know on your SAT date.

Take the SAT More than Once

No matter how prepared you are for your SAT, you're bound to be a little nervous and uncertain your first time. But if you take the test more than once, you'll begin to know what to expect and be able to plan ahead for it.

So, even if most of your high school takes their SAT in May or June of their junior year, consider taking the test for the first time as early as you can, just for practice. That way, you know that your first time taking the test doesn't have to count, so you can relax and (dare I say?) have fun with it — or at least minimize your anxiety throughout the process.

And remember that, in most cases, you can take your final SAT in December of your senior year and still get your scores in time to apply for college the following September.

Chapter **22**

Ten Tips to Be at Your Best on the SAT

The SAT is a big deal, right? So for the 24 hours before the test, you're entitled to do whatever it is you do to gain peace, focus, and self-empowerment. If that sounds a bit selfish, so be it. It's only for one day, so your friends and family should understand, provided you don't act like a complete heel. Here are my top ten recommendations for the day and night leading up to the SAT.

Do Something Fun the Day Before the Test

The day before your SAT is *your day* to do whatever you like to feel good: Go out to dinner, order a pizza, spend time with friends, spend time alone, watch a romantic comedy, watch a scary movie, go swimming, go roller skating, play with the dog, go out to the mall — whatever makes you happy and glad to be alive.

This includes taking the day before the test off from work. I mean it. Your job is important, but the SAT is more important than any single day of work. Given proper notice (at least two weeks if possible), any employer should be ready, willing, and able to accommodate you.

Don't Study for More than 20 Minutes the Night Before the Test

A common piece of advice is not to study the night before the SAT. I recommend this, too, and if it works for you, then definitely use it. I understand, however, that some students simply won't be able to put their books and notes away and not look at them.

So if you're convinced that *not* studying the night before the test will make you more nervous than studying, here's a compromise: Study no more than 20 minutes the night before. Don't take a practice test or try to cram in any new information. Instead, just look over a few notes and reread what you already know to reinforce it.

Pack Everything You Need the Night Before

REMEMBER

I recommend that you make a checklist of everything you want to bring to the test, pack these items the night before, and place them by the door so you won't forget them. Here are some must-have items for the SAT:

» Admission ticket

» Photo ID

» Calculator with fresh batteries

» Extra batteries

» Lots of sharp number 2 pencils

Additionally, here are a few extra items you may consider:

» Water bottle

» Convenient snacks (energy bars or trail mix is perfect)

» Pocket pencil sharpener

» Tissues

Also, if you have any special-needs items, consider these as well. For example, if you wear contact lenses, you may want to bring an extra pair or a pair of glasses just in case.

Do Something Relaxing Before Bed

The night before your SAT is a good time to just take it easy and wind down so you can get to bed early. It's natural to be nervous, so do whatever you can to quiet yourself before bed so that you sleep better. I recommend soft lighting, a cup of hot milk or chamomile tea, and something really boring on TV, like a documentary on the history of the paper clip.

Get a Good Night's Sleep

TIP

Get to bed early and, if you have trouble sleeping, try this trick: Put your pillow at the foot of the bed and sleep in a different direction from your usual position. Sounds goofy, but it works every time!

Wear Several Layers of Clothing

Nothing is worse in my experience than being too hot or too cold. So wear enough layers of clothing that you can remove or add to them, depending upon the temperature of the room.

Arrive at the Test Site Extra Early

The last thing you want is to get a late start or run into an unexpected delay and arrive at the test site stressed and out of breath. So plan to get to the test site especially early.

By the way, be sure to check the date of your test. I have a friend who had some really tense moments when she arrived and the test administrators for a *different* test told her the SAT wasn't being given that day. She discovered that she was one week early for her test!

Spend Your Time Just Before the Test However You Please

Between your arrival at the test site and the start of the test, do whatever works for you.

Be alone if you want to be alone, talk to a friend if you feel like it, and excuse yourself from talking to anyone you don't care for or about anything you don't want to talk about. After the test is over, you can resume the normal give-and-take of social living. But until the test is over, give yourself full permission to take care of yourself the way *you* need instead of worrying about anybody else.

Remember to Breathe

Breathing is so important. I do it every day, and so should you.

Seriously, though, if you find yourself getting nervous as the test approaches or while you're taking it, stop and take a couple of slow, deep breaths. Whether you believe it or not, a bit of oxygen is very calming, provided you don't overdo it. And it's free, while supplies last. While you breathe, repeat to yourself the words, "I'm doing my best." You can do no better.

Skip Over Any Questions That Throw You

Read each question, but give yourself the liberty of skipping over any questions that you're really not sure what to do with. All the questions are scored equally, so you might as well start out by answering as many of the easy ones as you can.

After you've done as many easy and medium questions as you can (in both the multiple-choice and grid-in sections), feel free to circle back to the beginning and start working on the first question you jumped over. Compared to the tougher questions, an earlier question may not look quite as difficult as it did when you first encountered it.

Index

A

absolute value, 22–23

absolute value functions, 151, 152

ACT, 14–15

addition

  combining functions, 143–144

  of complex numbers, 248–249

  of fractions, 20

  with imaginary numbers, 247

  of numbers to functions, 140

  of radicals, 25–26

adjacent side, right triangle, 237

advanced math on SAT. *See also specific advanced topics*

  circles on *xy*-plane

    center-radius form, 252–254

    completing the square, 256

    identifying interior and exterior points, 255–256

    most common topics on SAT, 326

    overview, 251–252

    plotting points on circle, 254–255

  complex numbers

    adding and subtracting, 248–249

    dividing, 250–251

    most common topics on SAT, 326

    multiplying, 249–250

    overview, 19, 248

  imaginary number *i*

    calculating with, 246–248

    imagining and defining, 246

    most common topics on SAT, 326

    overview, 19, 245–246

  overview, 10–11, 245

algebra. *See also specific algebra concepts*

  Algebra I reboot overview, 39–40, 53, 67–68

  and arithmetic, 27–28

basic equations, solving, 54–55

difficult equations, solving

  overview, 58

  quadratic equations, 58–62

  rational equations, 62–64

equations, identities, and inequalities, 28

evaluating expressions, 40

expressions overview, 28–29

factoring expressions

  cubic expressions, 50–51

  difference of squares, 44–47

  greatest common factor, 43–44

  overview, 43

  quadratic trinomials, 48–50

  to solve for variables, 57

  to solve quadratics, 59–60

  sum and difference of cubes, 47

  three-term quadratic equations, solving, 59–60

  two-term quadratic equations, solving, 59

  when finding roots of quadratic function, 201–202

importance to SAT success, 2

inequalities, solving, 64–66

intermediate equations, solving, 55–58

overview, 10

polynomial basics, 29–30

simplifying expressions

  combining like terms, 41

  distributing to remove parentheses, 41–42

  FOILing to multiply contents of parentheses, 42–43

  overview, 40–41

terminology, understanding, 27–30

anchor point (locator point) of function, 149

Angela's flowers word problem, 84

angles
  corresponding, 228–229
  linear pairs, 227–228
  overview, 227
  radian measure for, 242–244
  in triangles, 229
  vertical, 227–228
answer sheets, for practice tests
  practice test 1, 260
  practice test 2, 290
answers, on SAT
  answering every question, 12–13
  filling into grid, 8–10
  guessing on, 11–12
  number of questions to answer, 13–14
  practice SAT math tests, 273–287, 305–319
arc length, 242, 244
arithmetic, algebra and, 27–28. *See also*
    numbers; *specific operations*
average, mean
  calculating, 120, 130
  dot plots, 133
  most common topics on SAT, 325
  outliers, effect on, 121
axes, *xy*-plane, 31
axis of symmetry
  calculating, 182–183
  sketching parabola from standard-form
    quadratic functions, 184–185

B

balance scale method, 54
bar graphs, 131
bell-curve distribution, 121–122
bias, sample, 122–123
binomials
  defined, 30
  FOILing pair of, 42–43
  squares of, 196–197
"bounce" *x*-intercepts, sketching graph of,
    173–175
budgeting time on practice tests, 329

C

calculations
  doing basic in your head, 328
  statistical
    calculating mean and median of data set,
      120–121
    defined, 120
    finding minimum, maximum, and range, 120
    normal distribution of data sets, 121–122
    ordering data sets, 120
    outliers, 121
    overview, 120
    standard deviation, 122
Calculator section of SAT
  overview, 8
  practice test 1 answers and explanations,
    277–287, 310–319
  practice test 1 questions, 265–271, 296–303
calculators
  advanced use of, 34–35
  choosing, 33–34
  improving skill with, 328
  knowing basic use of, 34
  knowing which to avoid on SAT, 33
  overview, 33
cell phone calculators, 33
center-radius form, circles in, 252–254
charts, solving word problems with, 92–93
Cheat Sheet, explained, 3
child functions, 149
circles
  radian measure for angles, 242–243
  on *xy*-plane
    center-radius form, 252–254
    completing the square, 256
    identifying interior and exterior points,
      255–256
    most common topics on SAT, 326
    overview, 251–252
    plotting points on circle, 254–255
circumference, 242
class, studying for SAT in, 327

transformations of functions
  overview, 154
  stretch-compress-flip, 154, 158–160
  vertex form quadratic functions, 186–191
  vertical and horizontal, 154–158
  working with all three types of, 160–161
triangles. *See also* trigonometry
  angles in, 229
  building from trig ratios, 239–240
  congruent, 231
  isosceles, 229–230
  most important SAT identity, 241–242
  right
    defined, 231
    overview, 231–232
    Pythagorean Theorem, 232
    Pythagorean Triples, 232–234
    sines, cosines, and tangents, 237–238
    special, 234–237, 240–241
  similar, 111–112, 231
  sines, cosines, and tangents, 237–238
trigonometric functions, 151, 152, 153
trigonometry
  applying trig ratios to special right triangles, 240–241
  building triangles from single trig ratios, 239–240
  defined, 237
  most common topics on SAT, 325
  most important SAT identity, 241–242, 325
  overview, 225, 237
  radian measure and arc length, 242–244
  sines, cosines, and tangents, 237–238
trinomials
  defined, 30
  quadratic, factoring, 48–50
two-term quadratic equations, solving, 58–59
two-way tables, in probability problems, 126–128

U

unit circle, 242–243
unit conversion problems, 109–110
university requirements for SAT, 14–15

V

variables
  in algebra, 27–28
  factoring to solve for, 57
  isolating
    in basic algebra equations, 54
    in equations with multiple variables, 56
    factoring to, 57
    two-term quadratic equations, 58
  plugging into functions, 142–143
  solving equations with more than one, 55–58
  solving for expression containing multiple, 57–58
  stretch-compress-flip transformations, 189–191
  in terms, 29
  vertical and horizontal transformations, 186–189
vertex, quadratic functions
  combining vertical and horizontal transformations, 188–189
  finding, 183–184
  sketching parabola from standard form, 184–185
vertex form quadratic functions
  changing standard form to, 195–198
  changing to standard form, 194–195
  connecting standard forms and, 194–198
  overview, 186
  sketching parabola from, 191–194
  stretch-compress-flip transformations, 189–191
  vertical and horizontal transformations, 186–189
vertical angles, 227–228
vertical line test for functions, 147–149
vertical transformations
  combining with horizontal transformations, 157
  distinguishing horizontal transformations and, 156–157
  overview, 154–155
  vertex form quadratic functions, 186–189

About the Author

Mark Zegarelli is the author of *Basic Math and Pre-Algebra For Dummies* (John Wiley & Sons) plus nine other For Dummies books on math, logic, and test prep. He holds degrees in both English and math from Rutgers University and is a math, English, and test prep teacher and tutor.

Dedication

For Guang Guang, 我最喜欢的朋友。